MOLECULAR STRUCTURE AND LIFE

MOLECULAR STRUCTURE AND LIFE

Molecular Recognition of Nucleic Acids

Edited by
Yoshimasa Kyogoku and Yosifumi Nishimura

JAPAN SCIENTIFIC SOCIETIES PRESS Tokyo
CRC PRESS Boca Raton Ann Arbor London Tokyo

Supported in part by the Ministry of Education, Science and Culture under Grant-in-Aid for Publication of Scienctific Research Result.

Published jointly by:
JAPAN SCIENTIFIC SOCIETIES PRESS
2-10 Hongo, 6-chome, Bunkyo-ku, Tokyo 113, Japan
ISBN 4-7622-6719-8

and

CRC PRESS
2000 Corporate Blvd., N.W., Boca Raton, FL 33431, U.S.A.
ISBN 0-8493-7768-4

CIP Data available from the Library of Congress

Distributed in all areas outside Japan and Asia between Pakistan and Korea by CRC Press.

Printed in Japan

Preface

Recent progress in molecular biology reveals that many important biological processes such as embryogenesis, differentiation, and immunity are precisely controlled by a number of factors that are encoded in DNA. The most critical ones are considered to be transcription factors which regulate the expression of specific genes in each proccess. Even in a simple bacterial cell genes are strictly regulated by activators or repressors, responding to the environmental changes of the cell, most of which contain a well established DNA-binding motif, helix-turn-helix. In *Drosophila* many factors containing homeodomain are known to be responsible for the embryogenesis. In the differentiation of cell transcription factors are also found to regulate the fate of each cell. In addition, some protooncogene products, each of which is a cellular counterpart of a viral oncogene product, have been identified as transcription factors related to the differentiation of cells. In principle, the fate of each cell in a living system should be explained by the molecular recognition of DNA by these factors.

To understand the molecular mechanism of these biological processes we need to know the three dimensional structures of these transcription factors. These have now been investigated in detail: at the molecular structure level by a rapidly-developed spectroscopic method, by nuclear magnetic resonance (NMR) as well as by X-ray crystallographic analysis.

To emphasize the role of structural varieties of DNA and proteins in gene regulation, the International Symposium on Molecular Structure and Life: Molecular Recognition of Nucleic Acids was held December 10–14, 1991 at PACIFICO Yokohama as the 8th Yokohama Forum for the 21st century. Sponsors were Yokohama City, the Graduate School of Integrated Science of Yokohama City University and the Kihara Memorial Yokohama Foundation for the Promotion of Life Science in cooperation with The Spectroscopical Society of Japan, The Protein Engineering Society of Japan and with Grant-in-Aid for Scientific Research on Priority Areas, DNA binding proteins, from the Ministry of Education, Science and Culture of Japan. Many interesting lectures were presented in the symposium and several of the topics covered are summarized in this book.

A DNA molecule is not a simple featureless rod but takes a characteristic conformation depending on its base sequence as well as on its molecular environment. Each feature is recognized by a specific protein or an interacting drug. In addition, DNA containing a specific base sequence can bend with or without proteins. This bending character plays an important role in the regulation of some genes.

Transcription factors themselves show a variety of structures in their DNA binding domains, and newly-proposed structural motifs are also discussed. Several interaction modes between DNA and DNA-binding domains are described. Although structural alteration of both proteins and DNA induced by complexation makes precise prediction of the interaction site difficult, such interaction modes can be understood in a unified manner: specific nucleic acids bases are recognized by specific amino acid side chains through hydrogen-bonding, electrostatic or van der Waals interactions.

RNA polymerase can interact with various factors through protein-protein interaction on DNA. In eukaryotes genes are regulated in a complex structure, chromatin, in which a part of the DNA is methylated or super-twisted. This kind of structural change in chromatin should alter the level of gene regulation.

Finally, the bending ability of DNA is shown to play a role in DNA looping by protein-protein interaction. With these topics not only the interaction between DNA and protein but also the protein-protein interaction on DNA is presented as a critical point in gene regulation.

Based on the detailed molecular structures of DNA and proteins found in this book, we will know the molecular mechanism of some of the important biological processes, and in the near future we will be able to design such processes in living organisms or rescue cells from illness.

Yoshimasa KYOGOKU
Osaka University
Yoshifumi NISHIMURA
Yokohama City University

Contents

DNA-Binding Proteins and Their Interactions

Molecular Structure in Gene Regulation

DNA STRUCTURE AND CURVATURE

Molecular Structure and Life (Ed. by Y. Kyogoku and Y. Nishimura), pp. 3–27, Japan Sci. Soc. Press, Tokyo/CRC Press, Boca Raton (1992)

Understanding DNA Interactions and Deformations from 3-Dimensional Structures

Loren Dean WILLIAMS and Alexander RICH

*Department of Biology, Massachusetts Institute of Technology**

Even before Watson and Crick proposed that DNA forms a double helix, it was understood that DNA conformation is polymorphic. We now know that DNA can exist in two different right-handed double helical forms, A-DNA and B-DNA, and a left-handed form, Z-DNA. In addition, DNA can be bent, twisted and unstacked by small molecules such as ions, intercalators, and groove binders (some of these molecules are shown in Fig. 1), and macromolecules, such as proteins and nucleic acids. Both the major and minor grooves of DNA comprise binding sites for ions, drugs, and macromolecules. Without attempting to exhaustively review the literature, we describe the results of single crystal X-ray diffraction studies of non-covalent DNA complexes with ions, intercalators and groove binders.

I. CATIONS

Cations stabilize DNA against thermal denaturation (*1–4*), condense

*Cambridge, MA 02139, U.S.A.

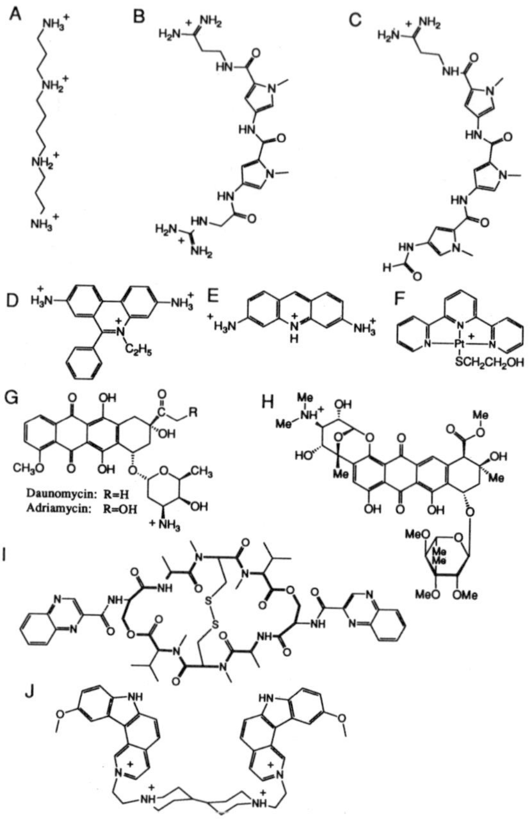

Fig. 1. DNA binding agents: (A) spermine, (B) netropsin, (C) distamycin, (D) ethidium, (E) proflavine, (F) TPH, (G) daunomycin/adriamycin, (H) nogalamycin, (I) triostin A, and (J) ditercalinium.

DNA into compact structures (*4–8*), and promote the B-DNA to Z-DNA transition (*9, 10*).

1. Sodium

Identification of a sodium ion in a sea of water is problematic. The X-ray data obtained from even the best diffracting DNA crystals is not of sufficient quality to observe hydrogen atoms. Thus ions are distinguished from water molecules only by the geometry of their coordination. In principle, the coordination sphere of a sodium ion is octahedral and the coordination sphere of a water molecule is tetrahedral. Facile distortion of the coordination spheres of both sodium ions and water molecules, including formation of two- and three-centered hydrogen bonds (*11*, *12*), makes it difficult in practice to distinguish between a sodium ion and a water molecule. The conservative approach is to assume that solvent peaks, even those forming distorted octahedra, are water molecules. Nonetheless, in high resolution structures, sodium ions can be identified with acceptable confidence. Sodium ions were located bound to Z-DNA crystallized with sodium as the lone cation at a resolution of 1.4 Å (*13*) and in certain B-DNA anthracycline complexes solved at resolutions ranging from 1.2–1.7 Å (*14*, *15*). It appears that sodium ions do not bind site-specifically to DNA and are structurally interchangeable with other monovalent (or even divalent) cations. In general, sodium ions do not induce specific conformational deformation of DNA.

However, in an outstanding exception, solution studies have shown that telomer sequence DNA does bind sodium in a site-specific manner. Telomer DNA forms G-quartets in a monovalent cation dependent manner, with stability; sodium > potassium, cesium > lithium (*16*). It was proposed by Williamson and coworkers that four carbonyl oxygens of a G-quartet (or eight carbonyl oxygens of two G-quartets) chelate a sodium ion in a manner reminiscent of a crown ether. Crystallographic support for this model is provided by the atomic resolution structure of the sodium salt of $[\mathrm{r}(AU)]_2$ (*17*). In this abbreviated duplex, two carbonyl oxygens (of the uracils) chelate a sodium ion in an analogous fashion to that proposed for G-quartets.

2. *Magnesium*

Magnesium forms a more regular and compact octahedron than does sodium, and can be identified with greater confidence. Magnesium ions have been observed bound to Z-DNA (*18*), a DNA-anthracycline complex (*19*), a DNA echinomycin complex (*20*), and yeast-tRNA^{Phe} (*21*). Magnesium can induce specific conformational changes in Z-DNA, from ZI to ZII (Fig. 2) (*18*). However, the observation of partial occupancy of the ZII conformation in Z-DNA crystals grown in the absence of magnesium (*22*) suggests that a sodium ion or possibly a water molecule can substitute for a magnesium ion in stabilizing the ZII conformation of Z-DNA.

Four distinct magnesium ions were located in the electron density maps of the orthorhombic form of yeast-tRNA^{Phe} at a nominal resolution 2.5 Å (*21*). Consistent with solution studies, analysis of the structure indicates that specific coordination of magnesium is required for proper folding of yeast-tRNA^{Phe}. For example, the sharp bend at the beginning

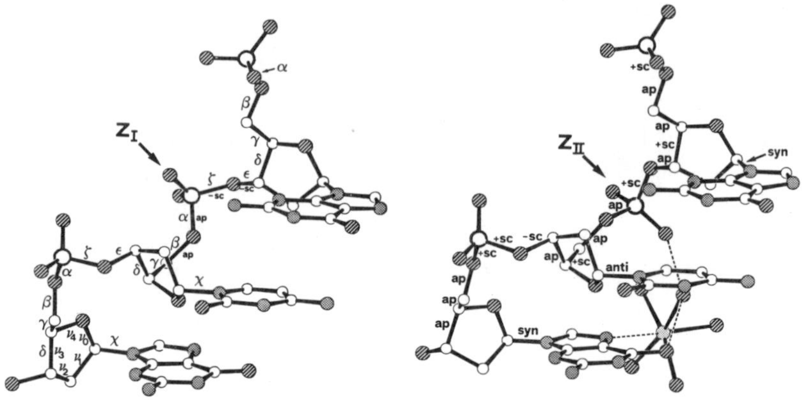

Fig. 2. Backbone conformations of Z-DNA; (A) ZI, and (B) ZII conformation. Oxygen atoms, both of the DNA and of water molecules, are striped, nitrogen atoms are stippled, and the magnesium bound to the ZII conformation (right) is also stippled (Reprinted from (*18*)).

of the D stem brings phosphates from different segments of the tRNA molecule into close proximity. Repulsive interactions are neutralized by magnesium ions and also by spermine molecules as discussed below.

3. Cobalt and ruthenium hexamine

Cobalt hexamine is five orders of magnitude more effective in stabilizing Z-DNA than sodium and four orders of magnitude more effective than magnesium (9). The crystal structure a cobalt hexamine complex of $[d(CGCGCG)]_2$, was solved to 1.3 Å resolution (*23*). This trivalent cation forms hydrogen bonds to the phosphate oxygens and to N7 and O6 positions of guanines on the convex surface. Ruthenium hexamine, with the same charge but a larger radius than cobalt hexamine, is much less effective in stabilizing Z-DNA (*24*). The crystal structure, a ruthenium hexamine complex of $[d(CGCGCG)]_2$, was solved to 1.2 Å resolution (*24*). One mode of ruthenium hexamine binding is nearly identical to that of cobalt hexamine. In a second mode, ruthenium hexamine bridges from the convex surface of one duplex to the edges of the deep groove of an adjacent duplex.

4. Polyamines

Spermine (Fig. 1A) is the largest and most complex polyamine found in eucaryotes. It has been observed by X-ray diffraction in complexes in several types of nucleic acid complexes (for a review of X-ray crystal structures of spermine-nucleic acid complexes, see (*25*)). Spermine has been observed bound to yeast-tRNA^{phe} (*21*), A-DNA (*26*), B-DNA anthracycline complexes (*15*), and Z-DNA (*18*, *22*). A glaring omission from this list is B-DNA. A spermine molecule was tentatively identified bound to a B-DNA dodecamer (*27*) but apparently did not refine in a well-behaved manner. Other polyamines, such as spermidine (smaller than spermine) and thermospermine (larger), have been observed by x-ray diffraction bound to Z-DNA (*28*, *29*) at 1.0 Å resolution.

We have recently reported (*22*) that spermine uniquely alters interactions between Z-DNA duplexes. In the presence of spermine plus trace amounts of sodium, the Z-forming DNA fragment [d(CGCGCG)]$_2$ crystallizes in the "spermine form". In essentially any other cationic environment (*i.e.*, in the presence of magnesium alone, magnesium plus spermine, sodium alone, cobalt or ruthenium hexamine, *etc.*) this DNA fragment crystallizes in a "non-spermine form" (for example, see (*13*, *18*, *23*, *24*)). The inter-duplex interactions of Z-DNA are significantly different in the two forms. The simultaneous interaction of three duplexes with a single spermine molecule in the spermine form of Z-DNA is shown in Fig. 3.

The conformation of Z-DNA in the spermine form is significantly different from the non-spermine form. In a conformational adjustment known as X-displacement, the base pairs are shifted around 0.5 Å away

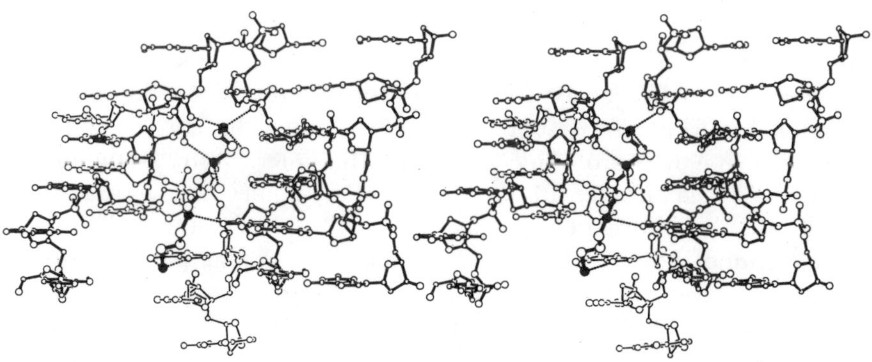

Fig. 3. ORTEP stereoview of a spermine molecule interacting simultaneously with three Z-DNA duplexes. Spermine bonds are solid, spermine carbons are drawn as circles with large radii and spermine nitrogens are stippled. One Z-DNA duplex is drawn with hollow bonds, single strands of two others are shown with solid thick bonds, and solid thin bonds. Hydrogen bonds are dashed and water molecules are drawn as circles with large radii (reprinted from (*22*)).

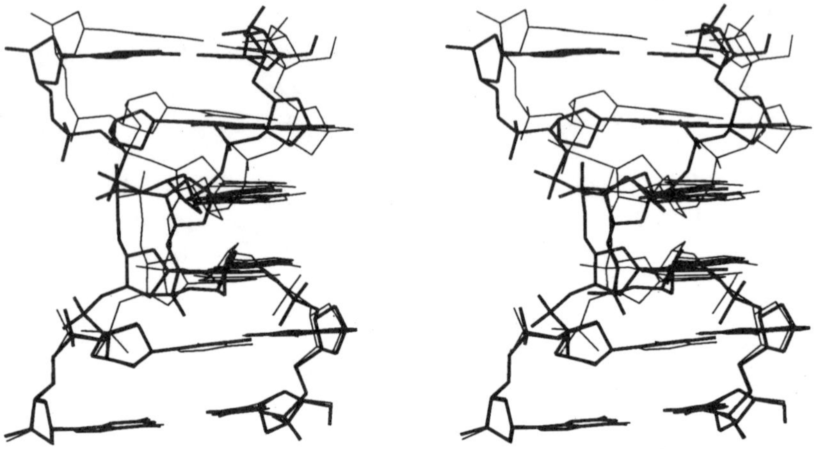

Fig. 4. Comparison of the conformations of $[d(CGCGCG)]_2$ in the pure-spermine form (bold lines) and in the magnesium form (thin lines). To emphasize the differences, the terminal residues of the two duplexes are superimposed (reprinted from (*22*)).

from the minor groove into the convex surface of the spermine form. In addition, the helix is shorter (Fig. 4) with an average helical rise per residue of 3.67 Å, over 0.1 Å less than in the non-spermine form. Thus a complete turn of a spermine form Z-DNA helix (12 base-pairs per turn) would be around 2 Å shorter than a complete turn of the non-spermine form. Although it is clear that Z-DNA has greater conformational freedom than earlier imagined, a full interpretation of Z-DNA deformations requires that the effects of ions be distinguished from those of the lattice. Two spermine molecules were located in the electron density maps of the orthorhombic form of yeast-$tRNA^{phe}$ (*21*). As in the case of Z-DNA, polyamines have a profound effect on the interactions between the yeast-$tRNA^{phe}$ molecules in the lattice. Polyamines are required for crystallization of diffraction quality yeast-$tRNA^{phe}$ crystals (*30*) and in this role cannot be replaced by other ions such as magnesium (also see (*31*)).

The effects of spermine on packing of nucleic acids in crystals suggests that the molecular basis for the tendency of spermine to stabilize compact nucleic acid structures derives in part from the capacity of spermine to interact simultaneously with several duplexes (or in the case of tRNA, several molecules). This capacity is maximized both by the polymorphic nature of the interactions (electrostatic, hydrogen bonding, and hydrophobic) of spermine and by the length (approximately 15 Å).

Clearly, spermine (other polyamines) mediate interactions between nucleic acid molecules in ways that are unlike effects of other cations. Although it is true that many of the properties of polyamines can be understood simply in terms of electrostatic interactions (between cationic polyamines and the anionic nucleic acids (*4*)), this model cannot account for the capacity of spermine to alter interactions between duplexes in a manner that is distinct from the effects of other cations.

II. GROOVE BINDING MOLECULES

The arc-shaped oligopeptides known as minor groove binders exhibit a wide range of antibiotic activities, including anti-bacterial, anti-fungal, anti-helminthic, and anti-viral. These cationic molecules act at the DNA level by binding preferentially in the minor groove of DNA containing AT base pairs (*32–34*). A series of complexes of minor groove binders bound to DNA dodecamers has been solved by X-ray diffraction. These include complexes of netropsin (*35*, *36*), distamycin (*37*), Hoechst 33258 (*38–40*), and 4′,6 -diamidino-2-phenylindole (DAPI, (*41*)). Netropsin is shown in Fig. 1B and distamycin in Fig. 1C.

1. *Netropsin*

X-ray structures of netropsin bound to several different DNA dodecamers have been solved at resolutions of around 2.2 Å (*35*, *36*). The drug fully occupies the minor groove of the DNA with extensive van der Waals contact between the drug and the walls and floor of the minor groove. The five membered pyrole rings are roughly parallel to the walls of the minor groove with the amide nitrogens forming hydrogen bonds to the electronegative atoms on the floor of the groove. The positively-charged ends of the drug molecule are located near the floors of the groove and do not form hydrogen bonds to negatively charged phosphate oxygens. The conformation of the DNA is very nearly the same as the conformation observed in the absence of netropsin. Similarly, the conformation of netropsin is nearly the same as observed in the absence of DNA (*i.e.*, in the crystal structure of the drug alone (*42*)).

2. *Distamycin*

The crystal structure of distamycin bound to the dodecamer [d(CGCA-AATTTGCG)]$_2$ has been solved at 2.2 Å resolution (*37*). Distamycin is similar to netropsin but contains an additional pyrole ring and lacks a positive charge at one terminus. The distamycin complex is shown in

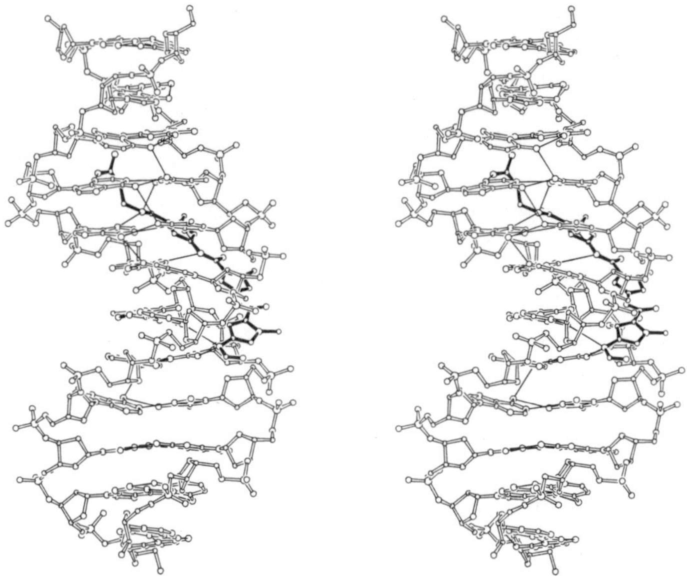

Fig. 5. ORTEP stereoview of the complex of distamycin bound to $[d(CGCAAATTTG\text{-}CG)]_2$. The bonds of the distamycin molecule are solid and the bonds to the DNA are open. Bifurcated hydrogen bonds that link the drug to the floor of the minor groove are shown as thin lines. Hydrogen bonds between bases, including the bifurcated interactions in the major groove associated with propellor twisting of the bases are also shown as thin lines (reprinted from (*37*)).

Figs. 5 and 6. Distamycin covers five base pairs of the minor groove while netropsin covers four. One unexpected feature of the distamycin complex was the propellor twisting of the DNA base pairs associated with bifurcated hydrogen bonds between successive AT base pairs in the A_3T_3 tract. However, this conformational variant of B-DNA was also observed in a related DNA fragment in the absence of a minor groove binding agent (*43*), and thus is not induced by distamycin.

The combined X-ray crystallographic results suggest that interaction of a minor groove binder with DNA does not induce significant conformational change in either the DNA or the drug. However, recent NMR evidence suggests that at high drug/DNA ratios, distamycin can interact with DNA in a second binding mode (*44*). In this mode, two drug molecules, stacked upon each other, simultaneously bind in the minor groove. This interaction, which has not yet been observed crystallographically, would require significant alteration of DNA conformation. In

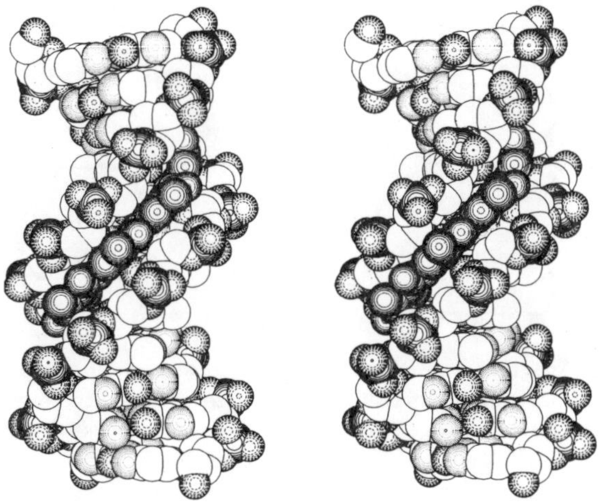

Fig. 6. Space filling stereoview, rotated 90° around the helical axis relative to the view in Fig. 5 of the complex of distamycin bound to $[d(CGCAAATTTGCG)]_2$. All atoms of the distamycin molecule are shaded with heavy black circles. In the DNA, nitrogen atoms are marked by dotted spheres, oxygen atoms as dashed spheres, phosphorus atoms as concentric circles, and carbon atoms are left unshaded (reprinted from (*37*)).

particular, the width of the minor groove would have to increase dramatically to accommodate two side-by-side distamycin molecules.

The specificity of minor groove binders for AT tracts of DNA is thought to derive in part from steric inhibition to binding at guanines (by the 2-amino group (*33*)) and in part from the 'snug fit' provided by an especially narrow minor groove of AT rich DNA. However, this supposition of 'snug fit' has been called into question by recent crystallographic evidence (with DNA decamers) that the minor groove of AT rich DNA is not necessarily more narrow than that of GC rich DNA (*45*).

III. INTERCALATORS

DNA intercalation, as first proposed by Lerman (*46*), is analogous to insertion of a 'false coin' (an intercalator) into a roll of pennies (base pairs). To intercalate in DNA, a planar molecule, generally with positive charge, slides between two intact base pairs. Intercalation extends the DNA helix and shifts the base pairs flanking the binding site in opposite directions along the helical axis (Fig. 7). Intercalators induce severe conformational changes in DNA (for recent reviews, see *47–49*). The diversity of the conformational states of DNA in intercalated complexes

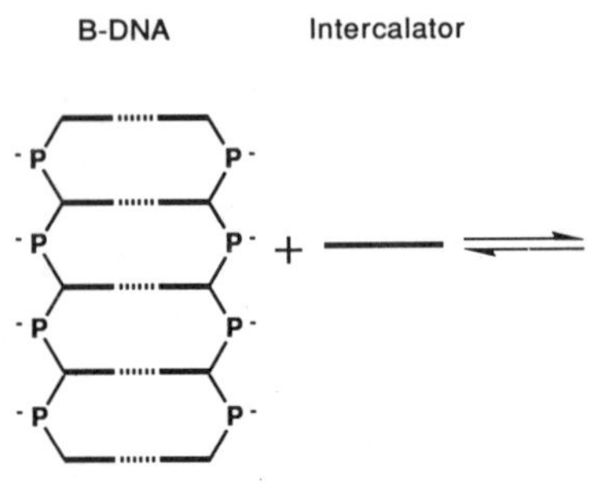

Fig. 7. Schematic diagram, viewed perpendicular to the helical axis, of the non-covalent reaction between DNA and an intercalator. The intercalator is represented by a long thick line, bases by short thick lines, Watson-Crick hydrogen bonds by dashed lines. The deoxyribo-phosphodiester backbone is represented by thin lines interrupted by P's (reprinted from (*47*)).

can be best understood in terms of 'induced fit'. The conformation of the DNA is dictated by the three dimensional shape of the intercalator as described in (*47*). However, the DNA conformation in intercalated complexes maintains certain B-like characteristics such as an orientation of the perpendiculars of the planes of the base pairs nearly parallel to the helical axis.

In the B-conformation of DNA, the helical twist is 36°. Each base pair is rotated around the helical axis by this amount relative to the preceding base pair. Intercalation invariably decreases the helical twist, unwinding the DNA in the vicinity of the binding site to less than 36° per base pair. The greatest net unwinding by a mono-intercalator is by ethidium with an unwinding angle of 26° per occupied intercalation site (*50*). In addition, DNA intercalation obeys the rule of neighbor-exclusion (*51*). Thus the two sites directly neighboring an occupied intercalation site must remain unoccupied or, in less absolute terms, intercalation is anti-cooperative at adjacent sites.

1. Simple intercalators—ethidium, proflavine, etc.

A series of structures of relatively simple intercalators bound to dinucleotides (both DNA and RNA) have been solved at atomic resolution (*52–57*). Ethidium is shown in Fig. 1D and proflavine in 1E. In each of these complexes the planar chromophore intercalates at a 5′ Pu-Py 3′ step. The complexes are underwound compared to B-DNA such that the long axes of the base pairs are nearly parallel to each other and to that of the intercalator. The result of this conformational deformation of the nucleic acid is nearly complete van der Waals contacts between the base pairs and the bound intercalator. With one exception, throughout the series of intercalated dinucleotides, the out-of-plane substituents of the intercalators (phenyls, ethyls, *etc.*) are located in the minor groove. The exception is the complex formed by the synthetic intercalator 2-hydroxyethanethiolato-2,′2,2″-terpyridine-platinium (II) (TPH, Fig. 1F) bound to the double helical DNA fragment $[d(CpG)]_2$ (*57*). The out-of-plane substituent of TPH is located in the major, not the minor, groove. Possible reasons for the unusual binding of TPH are discussed in the section on ditercalinium below.

2. Anthracyclines I—daunomycin, adriamycin, etc.

The anthracyclines are a family of widely used chemotherapeutic agents which inhibit DNA transcription *in vivo*. The activities of anthracyclines depend strongly on minor modifications in chemical structure. Daunomycin is used primarily for treating leukemia while adriamycin is used for solid tumors (*58*). Both are shown in Fig. 1G. A series of anthracycline I complexes (daunomycin, adriamycin, 4′-epiadriamycin, 11-deoxydaunomycin) have been crystallized and solved at resolutions ranging from 1.2 to 1.7 Å (*14*, *15*, *59–62*). In addition, the structure of a formaldehyde induced cross-link of daunomycin with DNA has been reported (*63*). The conformation of the DNA is nearly invariant throughout the series of complexes. In each complex, one anthracycline molecule intercalates at each of the two d(CG) steps of a hexamer duplex. The shape of these intercalators, with an out-of-plane substituent along the long axis of the intercalator, sterically prevents binding in the orientation observed in complexes with simple intercalators such as ethidium and TPH (above). Thus anthracyclines bind with their long axis perpendicular to those of the flanking base pairs. The complex of 4′-epiadriamycin bound to $[d(CGATCG)]_2$ (Fig. 8) illustrates the binding orientation of anthracyclines intercalated in DNA. In these complexes the DNA does not unwind at the intercalation site. However, the base pairs buckle, to wrap around the intercalator and maximize van der Waals contacts.

Although the orientation of the intercalated chromophore within the DNA is nearly invariant in this series, within the minor groove the orientation, conformation and DNA interactions of the amino-sugar are highly variable (Fig. 9). This variability is most pronounced in the complex which crystallized in a different crystal form from the other complexes (*62*). The variability in the crystalline state suggests that the amino-sugar may be mobile in the solution state.

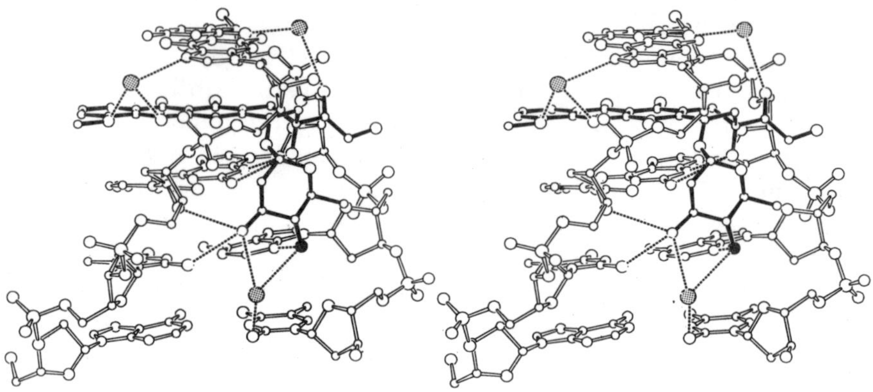

Fig. 8. ORTEP stereoview of the complex of 4′-epiadriamycin bound to $[d(CGATCG)]_2$. The DNA is drawn with hollow bonds, 4′-epiadriamycin is drawn with solid bonds and the hydrogen bonds linking the drug and the DNA are drawn with dashed lines. Water molecules, marked with stippling, are the largest spheres. The 4′-oxygen of 4′-epiadriamycin is marked by darker stippling (reprinted from (*61*)).

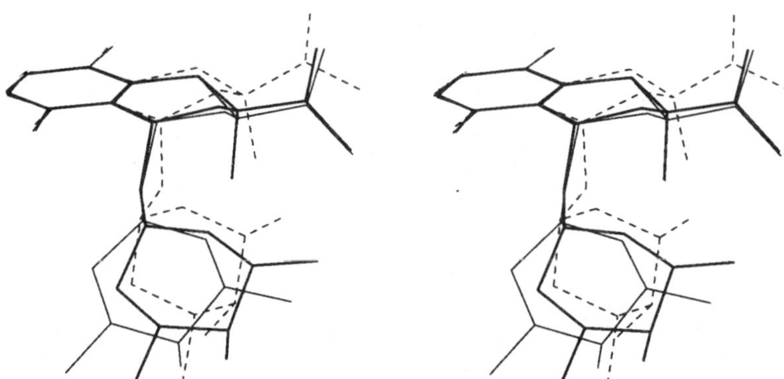

Fig. 9. Stereoview of the cyclohexene ring and amino sugar of daunomycin alone (dashed lines, (*90*)), in complex of 11-deoxydaunomycin with $[d(CGTsACG)]_2$ (thick lines), and in a complex of daunomycin with $[d(CGATCG)]_2$ (thin lines) (reprinted from (*62*)).

Throughout the series of crystallized DNA-anthracycline I complexes, neither the DNA nor the bound anthracycline appears to be conformationally stressed, confirming facile conversion to intercalator-induced DNA conformation.

3. *Anthracyclines II—nogalamycin*

Nogalamycin (Figs. 1H and 10) is a larger and more complex anthracy-

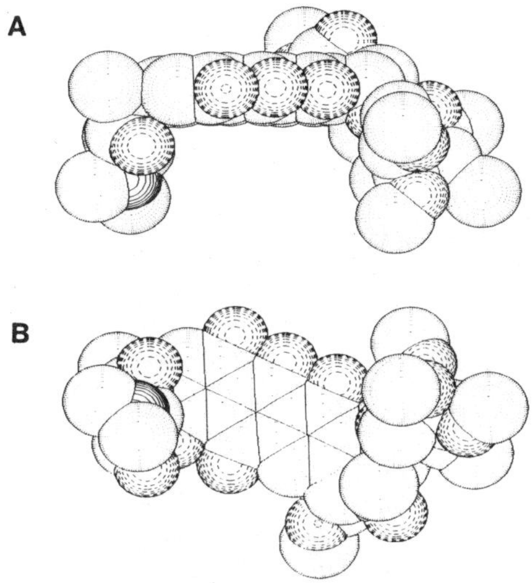

Fig. 10. Space filling representations of nogalamycin, viewed along the (A) edge, and (B) perpendicular to the plane of the chromophore. Carbon atoms are marked by concentric dots, oxygen atoms by concentric dashes, and nitrogen atoms by concentric lines (reprinted from (*47*)).

cline than daunomycin. Nogalamycin selectively inhibits transcription *in vivo* and is active against Gram-positive bacteria and experimental tumors. This drug is composed of a planar chromophore substituted with sugars on both ends and assumes the approximate shape of a dumbbell. On one end of the chromophore, nogalamycin contains both a methyl ester and a nogalose sugar, which is uncharged. The other end of the chromophore is fused to a bicyclo amino-sugar containing a positively charged dimethylamino group. Stimulated by the unusual dumbbell shape of nogalamycin, a number of investigators have shown that in spite of the obvious mechanistic obstacle, the drug forms a stable, intercalative complex with DNA. In addition to X-ray diffraction, intercalation of nogalamycin into DNA has been studied by modelling (*64*, *65*), NMR (*66*, *67*), and other solution experiments (*68*, *69*).

The slow association and dissociation rates (*69*) are consistent with the suggestion (*65*) that transient DNA melting is required for entry of nogalamycin into the duplex. We have proposed (*70*) an alternative mechanism of entry in which an opening to allow entry of nogalamycin could be created by unstacking and buckling of DNA base pairs without hydrogen bond disruption. In this mechanism, the DNA fluctuations would be of greater amplitude than required for entry of intercalators such as ethidium or daunomycin. In addition, it would conceivably be necessary to alter the conformation of nogalamycin to facilitate entry into the unstacked DNA.

Three-dimensional X-ray structures have been reported of several complexes of nogalamycin bound to the self-complementary DNA hexamer fragments, $[d(CGTsACG)]_2$ (*19*, *70–72*). The resolution of these structures ranges from 1.3 to 2.0 Å. In one complex the cytidines are methylated at the five-position although this appears not to affect the conformation of the complex. In each complex, one molecule of nogalamycin intercalates at each of the two CpG steps of the hexamer duplex.

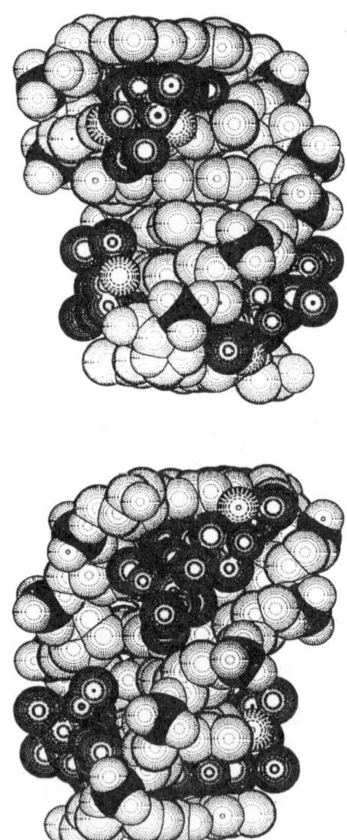

Fig. 11. Space filling representations of the complex formed by nogalamycin plus $[d(^{me}CGTsA^{me}CG)]_2$. (A) View into the major groove, and (B) view into the minor groove. The DNA is dotted except for the phosphorus atoms which are black. The atoms of nogalamycin are marked by concentric circles intersected with radial lines, except for the three oxygens which form direct hydrogen bonds to the DNA. These hydrogen-bonding oxygens are marked by dashed concentric circles.

Nogalamycin threads between the phosphodiester backbones and interacts with both strands in both grooves of the DNA (Fig. 11). In contrast to the DNA-anthracycline I complexes (where the positive charge is located in the minor groove), the positively charged amino-sugar of nogalamycin lies in the major groove. Similar to the DNA-anthracycline I complexes, the long axis of the chromophore of nogalamycin is roughly perpendicular to those of the flanking base pairs. The three aromatic rings are stacked within the DNA while the cyclohexene ring is completely unstacked and protrudes into the minor groove.

The DNA and the nogalamycin molecule are in close van der Waals contact, especially within the intercalative cavity and in the major groove. However, the three-dimensional shape of nogalamycin and the shape of the DNA are not a 'snug fit'. The long axis of nogalamycin exceeds the width of the DNA helix. To accommodate, both the DNA and nogalamycin appear to assume strained conformations in the complex. The base pairs buckle severely (by 26°) (Fig. 12). Even so, the potential for stabiliz-

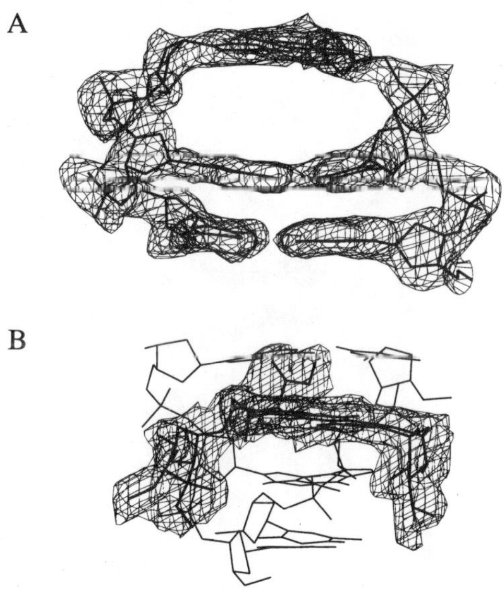

Fig. 12. Electron density sum maps (2Fo-Fc) surrounding the complex formed by nogalamycin plus $[d(^{me}CGTsA^{me}CG)]_2$. Electron density surrounding (A) the DNA alone, and (B) the nogalamycin molecule alone. The bonds of molecules around which electron density is shown are drawn with thick lines, bonds of the remaining parts of the complex are drawn with thin lines (reprinted from (*72*)).

ing van der Waals interactions is not fully realized. The uncharged nogalose sugar 'floats' in the minor groove, with only slight contact with the DNA. The poor fit could serve several biological functions. In a complex with a long fragment of DNA, helical distortion by nogalamycin could be transmitted for long distances down the helix, effecting interactions with other biological macromolecules. In addition, the poor fit serves to destabilize the complex and insures that the complex forms reversibly. The large surface area gives nogalamycin the potential to form extremely stable and potentially irreversible complexes with DNA. The poor fit insures reversible binding to DNA, a common and apparently effective mode of binding of biological anti-microbial agents.

IV. Bis-INTERCALATORS

1. Quinoxalines—triostin A, echinomycin

Triostin A (and echinomycin, which is distinguished by differences in the sulfur-containing cross bridge) is composed of two quinoxaline rings linked by a rigid arrangement of eight amino acids in a cyclic depsipeptide (Figs. 1I and 13). This molecule inhibits replication and transcription and is active against gram positive bacteria and experimental tumors.

A

B

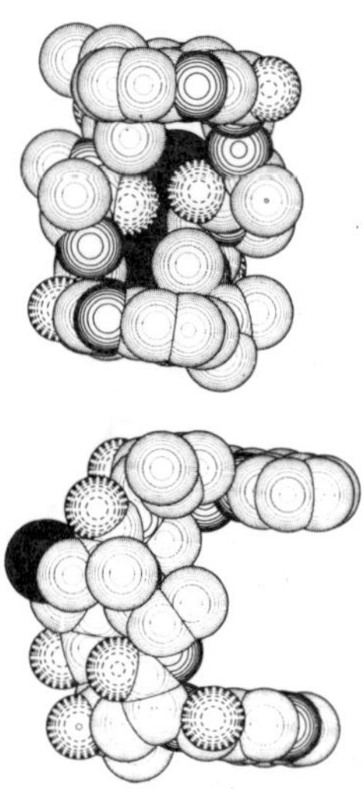

Fig. 13. Space filling representations of triostin A viewed (A) into the best plane of the linker, and (B) across the best plane of the linker. Carbon atoms are dotted, oxygen atoms are dashed, and nitrogen atoms are concentric circles. The sulfur atoms of triostin A are shaded.

Footprinting has shown that the linker is located within the minor groove and that the tightest binding of triostin A to DNA occurs at 5′ NCGN 3′ sites (where N = either A or T) (*73*, *74*). Three-dimensional X-ray structures of these two antibiotics have been solved in complexes with several fragments of DNA, the largest being $[d(CGCGTACGCG)]_2$ (*20*, *75*, *76*).

In the crystal structure, the quinoxaline rings bis-intercalate, and the cyclic depsipeptide is in close van der Waals contact with the minor groove of the DNA (Fig. 14). An unexpected feature of these complexes, solved at resolutions ranging from 1.7 to 2.2 Å, is the Hoogsteen interactions of some of the base pairs. However, the results of NMR experiments (*77*) indicate that in solution, the relevant conformation may involve open (unpaired) base pairs rather than Hoogsteen base pairs.

2. *Ditercalinium*

By combining rigid linkers to couple planar compounds, the synthetic

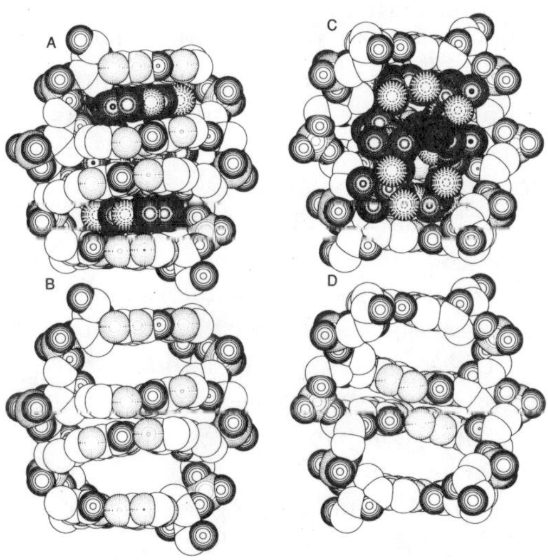

Fig. 14. Space filling representations of the triostin A-d(GCGT)-d(ACGC) complex. (A) The complete (*i.e.*, the asymmetric unit) complex viewing into the major groove, (B) the same view omitting the triostin A molecule, (C) the complete complex viewing into the minor groove, and (D) the same view omitting the triostin A molecule. In the DNA, carbon atoms are white, nitrogen atoms are dotted, oxygen atoms are lined, and phosphorus atoms are dashed. In the triostin A molecule, concentric circles with radial lines were used for all atoms except for the nitrogen atoms, which are dashed, and the phosphorous atoms, which are shaded.

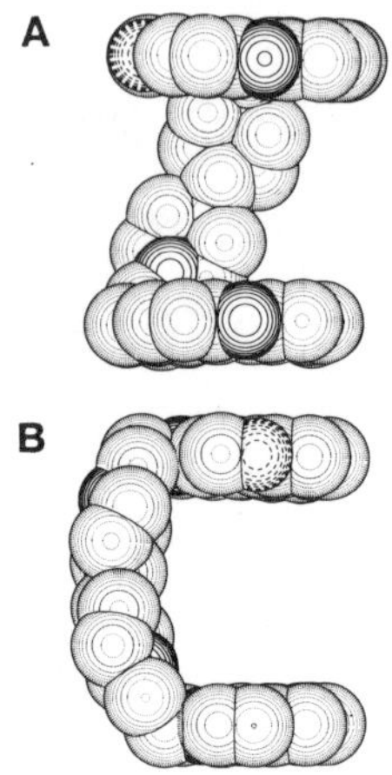

Fig. 15. Space filling representations of ditercalinium viewed (A) into the best plane of the linker, and (B) across the best plane of the linker.

7H-pyridocarbazole dimer, ditercalinium (Figs. 1J and 15), was designed/discovered by Roques, Le Pecq, and coworkers (*78*, *79*). Ditercalinium is unique among known DNA-binding chemotherapeutic agents, causing cell death by an unprecedented mechanism. Unlike most other DNA-binding chemotherapeutic agents, the mechanism of action of ditercalinium does not involve inhibition of DNA replication or transcription processes. Instead, DNA-ditercalinium complexes provoke malfunction of DNA repair systems (*80–82*).

As first shown by NMR (*83–87*), ditercalinium bis-intercalates in DNA, with the positively-charged linker of the drug located within the major groove. The DNA retains a right-handed, although underwound, double-helical conformation. We have recently described (*88*) a 1.7 Å resolution structure of a complex of ditercalinium bound to a double-stranded DNA fragment, $[d(CGCG)]_2$ (Fig. 16).

Surprisingly, from the minor groove, the ditercalinium complex mimics normal DNA. Both in shape of van der Waals surface and in arrangement of hydrogen bonding functionalities, within the minor groove the chromophores of ditercalinium simulate base pairs. Another striking feature of the complex is the discrimination of ditercalinium for guanines over cytosines. Both the intercalative and groove binding segments of ditercalinium interact almost exclusively with the guanines of the DNA fragment but not with the cytosines. The terminal cytosines of the DNA fragment are completely excluded from van der Waals contacts

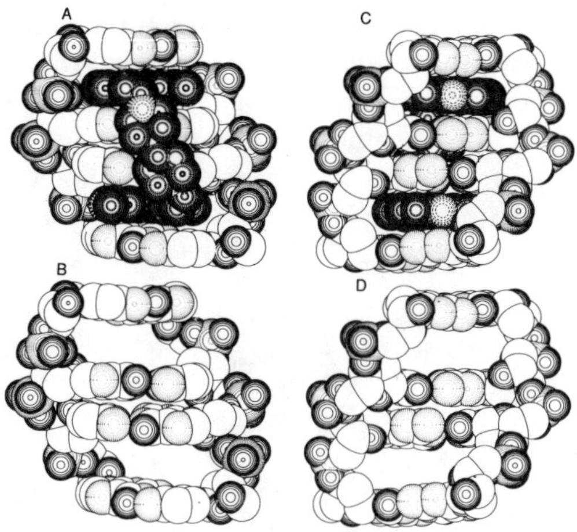

Fig. 16. Space filling representations of the ditercalinium-[d(CGCG)]$_2$ complex. (A) The complete complex viewing into the major groove, (B) the same view omitting the ditercalinium molecule, (C) the complete complex viewing into the minor groove, and (D) the same view omitting the ditercalinium molecule. In the DNA, carbon atoms are white, nitrogen atoms are dotted, oxygen atoms are lined, and phosphorus atoms are dashed. In the ditercalinium molecule, concentric circles with radial lines were used for all atoms except for the nitrogen atoms, which are dashed.

with the chromophores of ditercalinium and are 'unstacked' from the rest of the complex.

From inspection, it is not obvious why the linker of ditercalinium is located in the major rather than the minor groove. In nearly all other complexes of intercalators bound to fragments of DNA or RNA, out-of-plane substituents are located in the minor groove (nogalamycin occupies both grooves). From model building, it is clear that minor groove binding of ditercalinium is not blocked by steric interactions.

One clue to the unusual mode of binding of ditercalinium is provided by similarities with the mono-intercalator TPH (Fig. 1F, see section on simple intercalators above). TPH and ditercalinium are the only intercalators crystallized thus far with the minor groove unoccupied and with an out-of-plane substituent in the major groove. The structural feature that is common between TPH and ditercalinium, and is different from most other intercalators, is the large surface area of the intercalative chromophore. The chromophore of ditercalinium (four fused rings) and the chromophore of TPH (five fused rings) are significantly larger than the chromophores of other intercalators crystallized thus far (maximum three fused rings). It is reasonable to suggest that minor groove binding is favored over major groove binding by van der Waals interactions with the floor and walls of the minor groove. However, specific, directional interactions between the chromophore and the flanking base pairs can overcome this minor groove preference and direct the out-of-plane substituent to the major groove. These directional interactions between

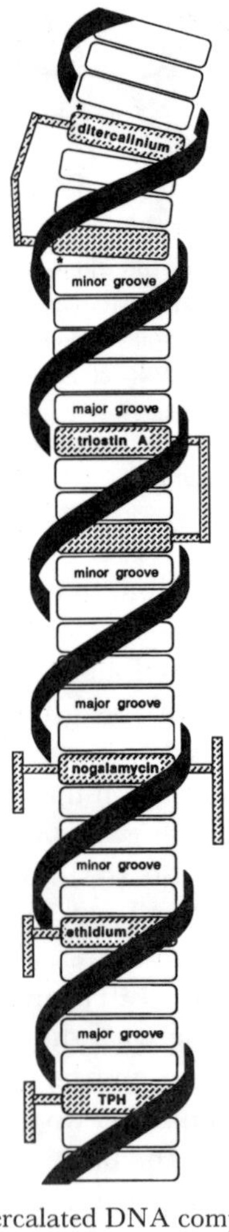

Fig. 17. Schematic diagram of intercalated DNA complexes. The intercalators, from top to bottom are ditercalinium, triostin A, nogalamycin, ethidium, and TPH. The drug molecules are shaded. The asterisks indicate sites of unstacking.

the chromophore and the flanking base pairs increase with the size of the chromophore.

SUMMARY

Structure and deformations of DNA and its complexes are critically related to biological function. We are just beginning to address such issues as (a) recognition of unperturbed DNA structure, (b) mimicry of unperturbed DNA, (c) deformations as in bending of the helical axis, and (d) effects of local instability, leading to increased deformability as caused by unstacking of bases. For a series of intercalated complexes solved by X-ray diffraction, each of these features is shown schematically in Fig. 17.

The influence of crystal packing on molecular conformation appears to be generally more pronounced with nucleic acids than with proteins. For example, crystal lattice-induced bending of DNA has been observed (*89*). Although DNA conformation in complexes of intercalators and groove binders may be less influenced than DNA alone due to the tendency of these agents to stabilize and stiffen DNA, lattice contacts may still select for a conformation that is not predominant in solution. The effects of lattice on a crystal structure can be determined by solution studies and by crystallizing a given fragment or complex in more than one crystal form. The lattice contacts and packing forces vary from one crystal form to another. If the DNA conformation in different crystallographic environments is unchanged, then the observed conformation is likely to be near an energy minimum in solution. Further, differences in conformation may reflect conformational heterogeneity in solution. By obtaining multiple crystal forms, we have recently analyzed effects of crystal packing on conformation and hydration of DNA-anthracycline complexes and on Z-DNA. It is clear that a complete understanding of complexes such as ditercalinium requires crystallization in additional crystal forms and with longer fragments of DNA.

Acknowledgments

This research was supported by grants from the National Institutes of Health, the American Cancer Society, the National Science Foundation, the Office of Naval Research and the National Aeronautics and Space Administration. L.D.W. acknowledges fellowship support by the National Institutes of Health and the Medical Foundation, Boston, MA.

REFERENCES

1. H.R. Mahler, B.D. Mehrotra and C.W. Sharp: *Biochem. Biophys. Res. Commun.* **4**, 79–82 (1961).
2. M. Mandel: *J. Mol. Biol.* **5**, 435–441 (1962).
3. H. Tabor: *Biochemistry* **1**, 496–501 (1962).
4. V.A. Bloomfield and R.W. Wilson: In "Polyamines in Biology and Medicine" (Ed. by D.R. Morris and L.J. Marton), pp. 183–206, Marcel Dekker, New York (1981).
5. L.C. Gosule and J.A. Schellman: *J. Mol. Biol.* **121**, 311–326 (1978).
6. R.W. Wilson and V.A. Bloomfield: *Biochemistry* **18**, 2192–2196 (1979).
7. J. Widom and R.L. Baldwin: *J. Mol. Biol.* **144**, 431–453 (1980).
8. D. Sen and D.M. Crothers: *Biochemistry* **25**, 1495–1503 (1986).
9. M.J. Behe and G. Felsenfeld: *Proc. Natl. Acad. Sci. U.S.A.* **78**, 1619–1623 (1981).
10. H.H. Chen, M.J. Behe, and D.C. Rau: *Nucl. Acids Res.* **12**, 2381–2389 (1984).
11. J.A. Jeffrey and H. Maluszynska: *Acta Cryst.* **B46**, 546–549 (1990).
12. R. Taylor, O. Kennard, and W. Versichel: *J. Am. Chem. Soc.* **106**, 244–248 (1984).
13. B. Chevrier, A.C. Dock, B. Hartmann, M. Leng, D. Moras, M.T. Thuong, and E. Westhof: *J. Mol. Biol.* **188**, 707–719 (1986).
14. A.H.-J. Wang, G. Ughetto, G.J. Quigley, and A. Rich: *Biochemistry* **26**, 1162–1163 (1987).
15. C.A. Frederick, L.D. Williams, G. Ughetto, G.A. van der Marel, J.H. van Boom, A. Rich, and A.H.-J. Wang: *Biochemistry* **29**, 2538–2549 (1990).
16. J.R. Williamson, M.K. Raghuraman, and T.R. Cech: *Cell* **59**, 871–880 (1989).
17. J.M. Rosenberg, N.C. Seeman, J.J.P. Kim, F.L. Suddath, H.B. Nichols, and A. Rich: *Nature* **243**, 150–154 (1973).
18. R.V. Gessner, C.A. Frederick, G.J. Quigley, A. Rich, and A.H.-J. Wang: *J. Biol. Chem.* **264**, 7921–7935 (1989).
19. Y.-G. Gao, Y.-C. Liaw, H. Robinson, and A.H.-J. Wang: *Biochemistry* **29**, 10307–10316 (1990).
20. G. Ughetto, A.H.-J. Wang, G.J. Quigley, G.A. van der Marel, J.H. van Boom, and A. Rich: *Nucl. Acids Res.* **13**, 2305–2323 (1985).
21. G.J. Quigley, M.M. Teeter, and A. Rich: *Proc. Natl. Acad. Sci. U.S.A.* **75**, 64–68 (1978).
22. M. Egli, L.D. Williams, Q. Gao, and A. Rich: *Biochemistry*, in press.
23. R.V. Gessner, G.J. Quigley, A.H.-J. Wang, G.A. van der Marel, J.H. van Boom, and A. Rich: *Biochemistry* **24**, 237–240 (1985).
24. P.S. Ho, C.A. Frederick, D. Saal, A.H.-J. Wang, and A. Rich: *J. Biomol. Struct. Dynamics* **4**, 521–534 (1987).
25. L.D. Williams, C.A. Frederick, R.V. Gessner, and A. Rich: In "Molecular Conformation and Biological Interactions" (Ed. by P. Balaram and S. Ramaseshan), pp. 295–309, Indian Academy of Sciences, Bangalore (1991).
26. S. Jain, G. Zon, and M. Sundaralingam: *Biochemistry* **28**, 2360–2364 (1989).
27. H.R. Drew and R.E. Dickerson: *J. Mol. Biol.* **151**, 535–556 (1981).
28. K. Tomita, T. Hakoshima, K. Inubushi, S. Kunisawa, H. Ohishi, G.A. van der Marel, J.H. van Boom, A.H.-J. Wang, and A. Rich: *J. Mol. Graphics* **7**, 71–75 (1989).
29. H. Ohishi, S. Kunisawa, G. van der Marel, J.H. van Boom, A. Rich, A.H.-J. Wang,

K. Tomita, and T. Hakoshima: *FEBS Lett.* **284**, 238-244 (1991).
30. S.H. Kim, G.J. Quigley, F.L. Suddath, and A. Rich: *Proc. Natl. Acad. Sci. U.S.A.* **68**, 841-845 (1971).
31. H. Prinz, N. Furgac, and F. Cramer: *Biochim. Biophys. Acta* **447**, 110-115 (1976).
32. C. Zimmer, K.E. Reinert, G. Luck, U. Wahnert, G. Lober, and H. Thrum: *J. Mol. Biol.* **58**, 329-348 (1971).
33. R.M. Wartell, J.E. Larson, and R.D. Wells: *J. Biol. Chem.* **249**, 6719-6731 (1974).
34. P. Dervan: *Science* **232**, 464-471 (1986).
35. M.L. Kopka, C. Yoon, D. Goodsell, P. Pjura, and R.E. Dickerson: *Proc. Natl. Acad. Sci. U.S.A.* **82**, 1376-1380 (1985).
36. M.L. Kopka, C. Yoon, D. Goodsell, P. Pjura, and R.E. Dickerson: *J. Mol. Biol.* **183**, 553-563 (1985).
37. M. Coll, C.A. Frederick, A.H.-J. Wang, and A. Rich: *Proc. Natl. Acad. Sci. U.S.A.* **84**, 8385-8389 (1987).
38. P.E. Pjura, K. Grzeskowiak, and R.E. Dickerson: *J. Mol. Biol.* **197**, 257-271 (1987).
39. M. Teng, N. Usman, C.A. Frederick, and A.H.-J. Wang: *Nucl. Acids Res.* **16**, 2671-2690 (1988).
40. M.A.A.F.C.T. Carrondo, M. Coll, J. Aymami, A.H.-J. Wang, G.A. van der Marel, J.H. van Boom, and A. Rich: *Biochemistry* **28**, 7849-7959 (1989).
41. T.A. Larsen, D.S. Goodsell, D. Cascio, K. Grzeskowiak, and R.E. Dickerson: *J. Biomol. Struct. Dynamics* **7**, 477-491 (1989).
42. H.M. Berman, S. Neidle, C. Zimmer, and H. Thrum: *Biochim. Biophys. Acta* **561**, 124-131 (1979).
43. H.C.M. Nelson, J.T. Finch, B.F. Luisi, and A. Klug: *Nature* **330**, 221-226 (1987).
44. J.G. Pelton and D.E. Wemmer: *Proc. Natl. Acad. Sci. U.S.A.* **86**, 5723-5727 (1989).
45. K. Yanagi, G.G. Prive, and R.E. Dickerson: *J. Mol. Biol.* **217**, 201-214 (1991).
46. L.S. Lerman: *J. Mol. Biol.* **3**, 18-30 (1961).
47. L.D. Williams, M. Egli, Q. Gao, and A. Rich: In "Structure and Function. I. Nucleic Acids" (Ed. by R.H. Sarma and M.H. Sarma), pp. 107-125, Adenine Press, Albany (1992).
48. W.A. Denny: *Anti-cancer Drug Design* **4**, 241-263 (1989).
49. S. Neidle, L.H. Pearl, and J.V. Skelly: *Biochem. J.* **243**, 1-13 (1987).
50. J.C. Wang: *J. Mol. Biol.* **89**, 783-801 (1974).
51. D.M. Crothers: *Biopolymers* **6**, 575-584 (1968).
52. H.M. Berman, W. Stallings, H.L. Carrell, J.P. Glusker, S. Neidle, G. Taylor, and A. Achari: *Biopolymers* **18**, 2405-2429 (1979).
53. K.K. Bhandary, T.D. Sakore, and H.M. Sobell: *J. Biomol. Struct. Dynamics* **1**, 1195-2010 (1984).
54. S.C. Jain and H.M. Sobell: *J. Biomol. Struct. Dynamics* **1**, 1179-1194 (1984).
55. S. Neidle, A. Achari, G.L. Taylor, H.M. Berman, H.L. Carrell, J.P. Glusker, and W.C. Stallings: *Nature* **269**, 304-307 (1977).
56. H.-S. Shieh, H.M. Berman, M. Dabrow, and S. Neidle: *Nucl. Acids Res.* **8**, 85-97 (1980).
57. A.H.-J. Wang, J. Nathans, G.A. van der Marel, J.H. van Boom, and A. Rich: *Nature* **276**, 471-474 (1978).
58. C.M. Haskell: "Cancer Treatment", W.B. Sanders Co., Philadelphia (1990).
59. G.J. Quigley, A.H.-J. Wang, G. Ughetto, G.A. van der Marel, J.H. van Boom, and A.

Rich: *Proc. Natl. Acad. Sci. U.S.A.* **77**, 7204–7208 (1980).
60. M.H. Moore, W.N. Hunter, B. Langlois d'Estaintot, and O. Kennard: *J. Mol. Biol.* **206**, 693–705 (1989).
61. L.D. Williams, C.A. Frederick, G. Ughetto, and A. Rich: *Nucl. Acids Res.* **18**, 5533–5541 (1990).
62. L.D. Williams, M. Egli, G. Ughetto, G.A. van der Marel, J.H. van Boom, G.J. Quigley, A.H.-J. Wang, A. Rich, and C.A. Frederick: *J. Mol. Biol.* **215**, 313–320 (1990).
63. A.H.-J. Wang, Y.-G. Gao, Y.-C. Liaw, and Y.-K. Li: *Biochemistry* **30**, 3812–3815 (1991).
64. S.K. Arora: *J. Am. Chem. Soc.* **105**, 1328–1332 (1983).
65. D.A. Collier, S. Neidle, and J.R. Brown: *Biochem. Pharmacol.* **33**, 2877–2880 (1984).
66. M.S. Searle, J.G. Hall, W.A. Denny, and L.P.G. Wakelin: *Biochemistry* **27**, 4340–4349 (1988).
67. X. Zhang and D.J. Patel: *Biochemistry* **29**, 9451–9466 (1990).
68. G.C. Das, S. Dasgupta, and N.N. Das Gupta: *Biochim. Biophys. Acta* **353**, 274–282 (1974).
69. J. Fok and M.J. Waring: *Mol. Pharmacol.* **8**, 65–74 (1972).
70. L.D. Williams, M. Egli, Q. Gao, P. Bash, G.A. van der Marel, J.H. van Boom, A. Rich, and C.A. Frederick: *Proc. Natl. Acad. Sci. U.S.A.* **87**, 2225–2229 (1990).
71. Y.-C. Liaw, Y.-G. Gao, H. Robinson, G.A. van der Marel, J.H. van Boom, and A.H.-J. Wang: *Biochemistry* **28**, 9913–9918 (1989).
72. M. Egli, L.D. Williams, C.A. Frederick, and A. Rich: *Biochemistry* **30**, 1364–1372 (1991).
73. P.B. Dervan and M.M. van Dyke: *Science* **225**, 1122–1127 (1984).
74. C.M.L. Low, H.R. Drew, and M.J. Waring: *Nucl. Acids Res.* **12**, 4865–4879 (1984).
75. A.H.-J. Wang, G. Ughetto, G.J. Quigley, T. Hakoshima, G.A. van der Marel, J.H. van Boom, and A. Rich: *Science* **225**, 1115–1121 (1984).
76. G.J. Quigley, G. Ughetto, G.A. van der Marel, J.H. van Boom, A.H.-J. Wang, and A. Rich: *Science* **232**, 1255–1258 (1986).
77. D.E. Gilbert, G.A. van der Marel, J.H. van Boom, and J. Feigon: *Proc. Natl. Acad. Sci. U.S.A.* **86**, 3006–3010 (1989).
78. D. Pelaprat, R. Oberlin, I. Le Guen, B.P. Roques, and J.B. Le Pecq: *J. Med. Chem.* **23**, 1330–1335 (1980).
79. D. Pelaprat, A. Delbarre, I. Le Guen, B.P. Roques, and J.B. Le Pecq: *J. Med. Chem.* **23**, 1336–1343 (1980).
80. B. Lambert, B.P. Roques, and J.B. Le Pecq: *Nucl. Acids Res.* **3**, 1063–1078 (1988).
81. B. Lambert, B.K. Jones, B.P. Roques, J.B. Le Pecq, and A.T. Yeung: *Proc. Natl. Acad. Sci. U.S.A.* **86**, 6557–6561 (1989).
82. B. Lambert, E. Segal-Bendirdjian, C. Esnault, J.B. Le Pecq, B.P. Roques, B. Jones, and A.T. Yeung: *Anti-Cancer Drug Design* **5**, 43–53 (1990).
83. A. Delbarre, M. Delepierre, J. Igolen, J.B. Le Pecq, and B.P. Roques: *Biochimie* **67**, 823–828 (1985).
84. A. Delbarre, M. Delepierre, B. Langlois D'Estaintot, J. Igolen, and B.P. Roques: *Biopolymers* **26**, 1001–1033 (1987).
85. A. Delbarre, M. Delepierre, C. Garbay, J. Igolen, J.B. Le Pecq, and B.P. Roques: *Proc. Natl. Acad. Sci. U.S.A.* **84**, 2155–2159 (1987).
86. M. Delepierre, J. Igolen, and B.P. Roques: *Biopolymers* **27**, 957–968 (1988).
87. M. Delepierre, R. Maroun, C. Garbay-Jaureguiberry, J. Igolen, and B.P. Roques: *J.*

Mol. Biol. **210**, 211–228 (1989).

88. Q. Gao, L.D. Williams, M. Egli, D. Rabinovich, S.-H. Chen, G.J. Quigley, and A. Rich: *Proc. Natl. Acad. Sci. U.S.A.* **88**, 2422–2426 (1991).
89. A.D. DiGabriele, M.R. Sanderson, and T.A. Steitz: *Proc. Natl. Acad. Sci. U.S.A.* **86**, 1816–1820 (1989).
90. S. Neidle and G. Taylor: *Biochim. Biophys. Acta* **479**, 450–459 (1977).

Molecular Structure and Life (Ed. by Y. Kyogoku and Y. Nishimura), pp. 29–42, Japan Sci. Soc. Press, Tokyo/CRC Press, Boca Raton (1992)

Effects of Base Pair and Sequence on DNA Structure and Drug-DNA Interaction

Seiichi UESUGI,*[1] Yasushi ODA,*[2] Hidekazu HIROAKI,*[3] Satoshi FUJII,*[3] Eiko OHTSUKA,*[4] Yasutoshi KAWASE,*[4] and Morio IKEHARA*[2]

Faculty of Engineering, Yokohama National University,[1] Protein Engineering Research Institute,*[2] Faculty of Pharmaceutical Sciences, Osaka University,*[3] and Faculty of Pharmaceutical Sciences, Hokkaido University*[4]*

DNA generally takes the B-form double helical structure in solution. However, the base sequence may affect the local structure producing some variation within the B-form. Accumulation of such variation in phase can even change the global shape of DNA molecules as observed in the case of DNA bending. In this paper, we present two examples of base sequence dependent change in conformation and properties of DNA. The first topic is the structure of I:A base pairs in DNA and the second is BLM(bleomycin)-DNA interactions.

I. EFFECT OF BASE SEQUENCES ON I:A BASE-PAIR STRUCTURE

We have studied mismatch base-pairing structures involving deoxyinosine (dI) in synthetic DNA oligonucleotides by NMR. It is known that hypoxanthine, which is the base of inosine and dI, behaves as a guanine analogue in nucleic acids, and is derived from adenine by treatment with

*[1] Yokohama 240, Japan
*[2] Suita, Osaka 565, Japan
*[3] Suita, Osaka 565, Japan
*[4] Sapporo 060, Japan

a mutagen, nitrous acid. DNA oligomers containing dI residues at positions corresponding to the third letters of the codons for a peptide sequence have been successfully used as hybridization probes (*1*, *2*). The basis for this method is that the dI residue is thought to make base pairs with multiple bases. It is assumed that an inosine residue in the wobble position of a tRNA anticodon can form base pairs with C, A, and U of mRNA codons (*3*).

1. Base Pairing Involving Deoxyinosine

Base pairing of dI with the four normal DNA bases, C, A, T, and G, was examined by ^{1}H NMR using d(G-G-I-X-C-C) and d(C-G-C-I-A-A-T-T-X-G-C-G) (X=C, A, T, and G) at neutral pH. The base-pair structures were identified by analysis of the temperature dependence of the chemical shifts and linewidths for the imino proton resonances, and nuclear Overhauser effects (NOE) between the exchangeable and/or nonexchangeable protons (*4*, *5*).

In d(G-G-I-X-C-C) duplexes, we identified dI(anti):dC(anti) and dI(anti):dA(anti) base pairs (*4*). We could not identify dI:dT and dI:dG base pairs because of the low stability of the duplex. In the d(C-G-C-I-A-A-T-T-X-G-C-G) duplexes, the dI formed stable base pairs with all four bases. The base-pair structures were assigned as dI(anti):dC(anti), dI(anti):dT(anti), dI(syn):dG(anti) (*5*), and dI(anti):dA(anti)/dI(anti):dA(syn) (Y. Oda *et al.*, in preparation) (Fig. 1). For the dI:dA pairing, dI(anti):dA(anti) and dI(anti):dA(syn) conformations coexist in solution of the d(C-G-C-I-A-A-T-T-A-G-C-G) duplex, whereas only the dI(anti):dA(syn) conformer is found in crystals (*6*), and only a dI(anti):dA(anti) conformer is found in solution of d(G-G-I-A-C-C) duplex. For the dI:dT base pairing, the same structure has been found in d(C-A-G-T-G-G-C):d(G-C-C-I-C-T-G) duplex (*7*) and in crystals of d(G-G-I-G-C-T-C-C) duplex (*8*).

These results confirm that a hypoxanthine base can form hydrogen-bonded base pairs with all four normal bases, C, A, T, and G, in DNA, proving that dI can be used as a universal base in hybridization probes.

2. Thermal Stability of DNA Duplexes Containing Deoxyinosine

The melting temperature (T_m) values of the d(C1-G2-C3-I4-A5-A6-T7-T8-X9-G10-C11-G12) duplexes were determined from the chemical shift-temperature profiles for the T7-CH3 proton resonance. T_m values were 61°C for X=dC, 55°C for X=dA, 51°C for X=dT, and 45°C for X=dG. The results indicate the order of stability to be: dI:dC>dI:dA>dI:dT>

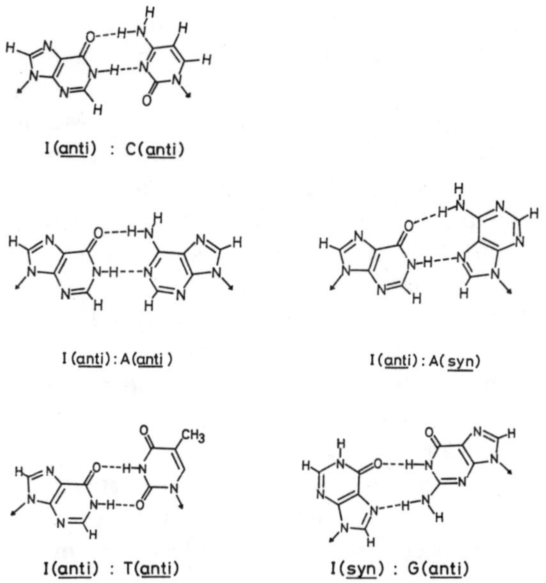

Fig. 1. Structures of the base pairs containing dI.

dI:dG. This is in accordance with the earlier results obtained by optical melting techniques (*9–11*).

3. Effect of Sequence on dI:dA Base-pair Structure

As described above, two types of dI:dA mismatch base-pair structures have been identified in the different base sequences: dI(anti):dA(anti) in the d(G-G-I-A-C-C) duplex (*4*) and dI(anti):dA(syn) in the d(C-G-C-I-A-A-T-T-A-G-C-G) (*5*). Similar phenomena have been observed for dG:dA mismatch pairing (*12–15*). The sequence-dependent environment may determine the conformation of these purine:purine mismatch pairs.

dI(anti):dA(anti) and dI(anti):dA(syn) conformations coexist and the dI(anti):dA(syn) conformer is predominant in d(C-G-C-I-A-A-T-T-A-G-C-G) at low temperature (below 0°C) (Fig. 2). In crystals, the predominant conformer, dI(anti):dA(syn), is apparently observed, although crystal forces may affect the mismatch base-pair structures.

What factors determine the preferable dI:dA mismatch base-pair structure in the different sequences ? Stacking interactions with neighboring base pairs may be one important factor. To investigate this point, molecular dynamics studies were made on the d(G-G-I-A-C-C) duplex (designated duplex I) and the d(C-G-C-I-A-A-T):d(A-T-T-A-G-C-G) fragment of d(C-G-C-I-A-A-T-T-A-G-C-G) duplex (duplex II) (S. Fujii *et al.*, unpublished results). The obtained conformational energy data indicate that dI(anti):dA(anti) is more stable than dI(anti):dA(syn) in duplex I, whereas the two conformers have almost identical stability in duplex II; (dA(syn) is slightly more stable than dA(anti)). These results are consistent with those of the NMR study. Figure 3 shows the stacking interactions involving the dI:dA base pairs. In duplex II, the base overlaps at CpI and TpA steps are very small. The weak stacking may increase the chance of opening the minor groove. This dynamic feature at the pyrimidine-purine steps of the d(C-I-A):d(T-A-G) sequence may facilitate altering the dA base from anti to syn.

It should be noted that hydration may also be an important factor, since this is a dominant feature stabilizing B-DNA helix (*14*). The width of the minor groove in the central -A-A-T-T- region of the d(C-G-C-I-A-A-T-T-A-G-C-G) duplex is very narrow (*6*). The structure of the AT-rich region is stabilized with a "spine of hydration" (*16*). The distinctive structure of the AT-rich region may affect the mismatch base-pair

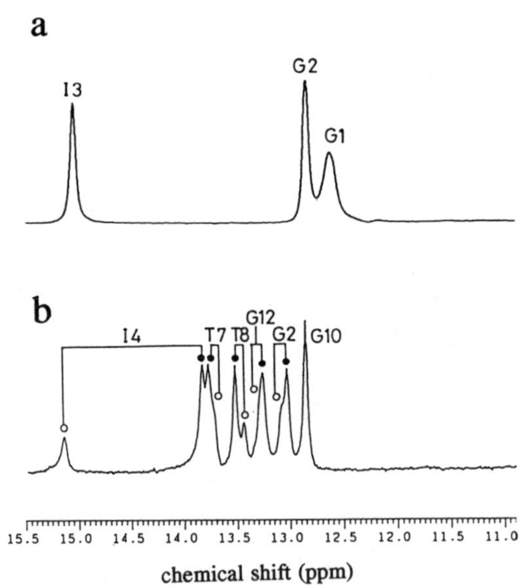

Fig. 2. ^{1}H-NMR spectra in the imino proton region in H_2O-D_2O (4 : 1) containing 0.1 M NaCl, 10 mM sodium phosphate buffer: (a) d(G1-G2-I3-A4-C5-C6) duplex at 1°C, pH 7.0, and (b) d(C1-G2-C3-I4-A5-A6-T7-T8-A9-G10-C11-G12) duplex at −10°C, pH 6.8. The imino proton resonances of d(C-G-C-I-A-A-T-T-A-G-C-G) duplex for dI(anti):dA(anti) and dI(anti):dA(syn) are designated by open circles and closed circles, respectively.

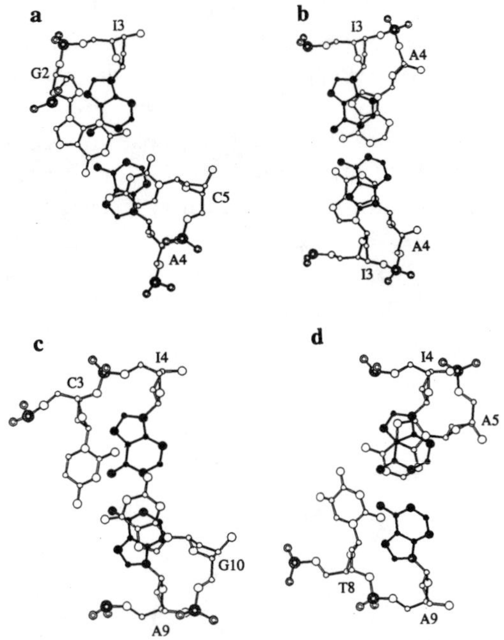

Fig. 3. Base stacking interactions involving the dI(anti):dA(anti) base pair of steps: (a) (G2:C5)/(I3:A4) and (b) (I3:A4)/(A4:I3) in the d(G1-G2-I3-A4-C5-C6) duplex, and (c) (C3:G12)/(I4:A11) and (d) (I4:A11)/(A5:T12) in the d(C1-G2-C3-I4-A5-A6-T7):d(A8-T9-T10-A11-G12-C13-G14) duplex.

structure: the relatively compact dI(anti):dA(syn) conformer is more stabilizing than the bulky dI(anti):dA(anti) conformer at the nearest-neighbor of the AT-rich region in the d(C-G-C-I-A-A-T-T-A-G-C-G) duplex.

The present studies on dI:dA pairing show a 'dynamic' feature of mismatch base pairing, which may be relevant to understanding mutation and its repair, in addition to the 'static' structural feature.

4. *Effect of dI-containing Mismatch Base Pairs on DNA*

The helical structure of the duplexes, d(G1-G2-I3-A4-C5-C6) and d(C1-G2-C3-I4-A5-A6-T7-T8-X9-G10-C11-G12) (X=dT, dG, and dA) were investigated by ^{1}H and ^{31}P NMR. The observed NOEs, coupling constants, and chemical shifts indicated these mismatch base pairs to be accommodated in the B-DNA duplex with only a subtle distortion of the local conformation.

For the bulky dI(anti):dA(anti) pair, the C1′-C1′ distance is 12.7 Å in the d(G-G-I-A-C-C) duplex, 2 Å wider than that of normal Watson-Crick base pair (*3*). The ^{31}P NMR spectra indicated some distortion of the backbone conformation probably caused by the greater distance between the paired residues (Fig. 4). The phosphorus resonance of G2pI3 showed a large downfield shift, suggesting that the phosphodiester torsion angle, ζ or α, is changed from normal *gauche*⁻ to *trans*. The altered conformation may correspond to the "BII" conformation observed in the d(C-C-A-A-G-A-T-T-G-G) duplex containing dG(anti):dA(anti) pair (*14*), and may allow expansion of the C1′-C1′ distance.

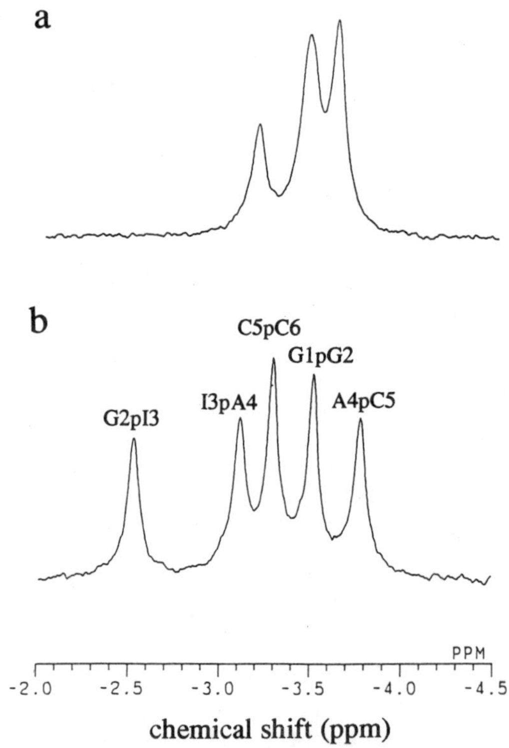

Fig. 4. ^{31}P-NMR spectra of (a) d(G1-G2-I3-C4-C5-C6) duplex and (b) d(G1-G2-I3-A4-C5-C6) duplex in H_2O-D_2O (4 : 1) containing 0.1 M NaCl, 10 mM sodium phosphate buffer (pH 7.0) at 1°C.

a

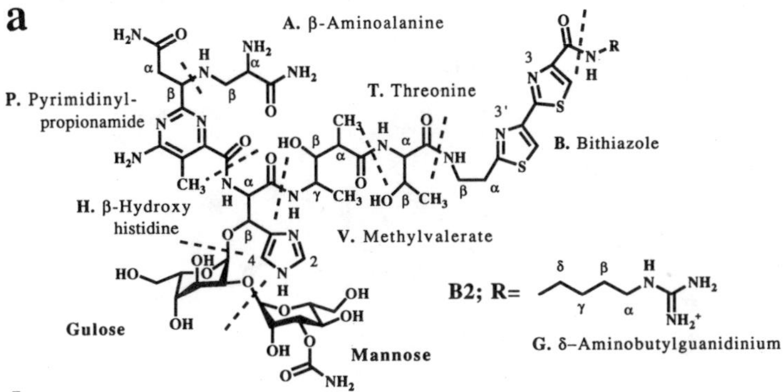

b

	1	2	3	4	5	6	7	8	9	10	11	12	
5'-	G	G	G	G	A	G	C	T	C	C	C	C	-3'
3'-	C	C	C	C	T	C	G	A	G	G	G	G	-5'
	12	11	10	9	8	7	6	5	4	3	2	1	

Fig. 5. Chemical structure of BLM-B2 (a) and structure and numbering system of d(G-G-G-G-A-G-C-T-C-C-C-C) (b). Abbreviations for the structural units in BLM are also shown.

II. EFFECT OF BASE SEQUENCES ON DNA RECOGNITION BY BLEOMYCIN

The BLMs (Fig. 5), antitumor glycopeptides produced by *Streptomyces verticillus*, are used clinically for certain carcinomas (*17*). The mechanism of cytotoxic action is due to their DNA degradation capability. This process proceeds catalytically in the presence of any of several redox active metal ions and oxygen, or ultraviolet light (*18*). Previous studies showed that BLM possesses two separate functional sites: a metal binding site which catalyzes activation of a molecular oxygen, and a DNA-binding site called 'tetrapeptide-S' with a terminal cationic residue (*19*). Thus, BLM is one of the 'micro-enzymes' which are a basis for the design of artificial 'chemical nucleases'.

The chemical reactions involved in the DNA degradation by BLM are understood, but the mode of the interaction and recognition between DNA and BLM remain unknown. BLM binds to double-stranded DNA and cleaves it in a sequence-specific manner at GpC and GpT (*18*). BLM has long been assumed to act as 'intercalator', and the bithiazole ring of BLM intercalates into base pairs of a double helix (*20*, *21*). On the other hand, a few groups proposed the 'minor-groove binding model' of BLM (*22*, *23*), a model more suitable for explaining many BLM observations. However, neither model has yet been proved because of lack of direct

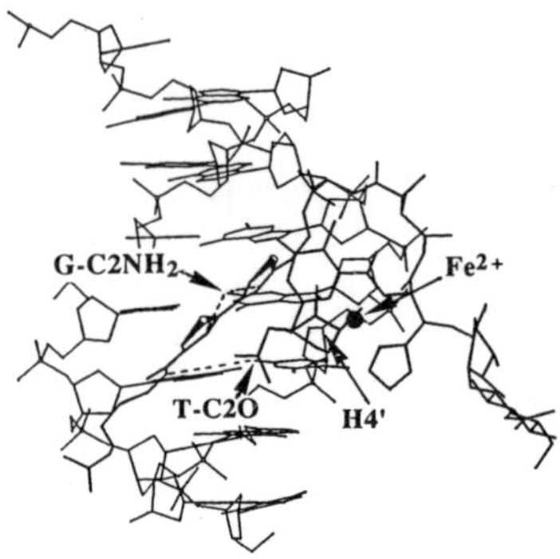

Fig. 6. Proposed binding model of Fe^{2+}-BLM-B2 to the half molecule of the d(C-C-A-C-C-T-A-G-G-T-G-G) duplex. The drug molecule is drawn in thick lines. The hydrogen-bonds between G9C2NH$_2$ and bithiazole N3/N3′ and between T10C2O and δNH of δ-aminobutylguanidinium residue are indicated by dashed lines.

evidence such as that obtained from X-ray crystallography or NMR spectroscopy.

Thus, we chose a simple strategy for solving this problem, use of an oligonucleotide system which contains a single preferred site for BLM cleavage. We have reported titration studies of d(C-C-C-C-A-G-C-T-G-G-G-G) duplex with 'inactive' BLM-metal complexes, BLM-Ni^{2+} and BLM-VO^{3+}, using ^{1}H and ^{31}P NMR (*24*). In our system, BLM-metal complexes specifically bind to the DNA at the site G6pC7, but no chemical shift changes for DNA proton and phosphorus resonances are observed. The results are not consistent with an 'intercalation model'.

Furthermore, in our studies using d(C-C-A-C-C-T-A-G-G-T-G-G) (*25, 26*), we observed specific upfield shifts of resonances for T10N3H of DNA and δNH of δ-aminobutylguanidinium residue of BLM, while the DNA helix structure was not perturbed upon BLM complex formation (*25*). We concluded that BLM recognizes the G9pT10 sequence from its minor-groove side with two hydrogen bonds, between G-C2NH$_2$ and bithiazole N3/N3′ and between T-C2O and δNH of the terminal cationic residue. A certain electrostatic interaction between the terminal cationic group of BLM and DNA phosphate backbone is also assumed. This model is basically identical to that proposed by Dickerson (*22*).

A molecular modeling study was done using the MOLGRAPH program (H. Hiroaki and S. Uesugi, in preparation), and a model obtained which can explain some features of BLM (Fig. 6). In this model, the reaction center of BLM is located very close to the H4′ of T10; BLM primarily attacks at H4′. The model can also explain that BLM cleaves the GpT sequence as well as the GpC sequence because both thymine and cytosine have a C2O carbonyl oxygen at the same position.

1. Altered Sequence Preference by Bleomycin of d(G-G-G-G-A-G-C-T-C-C-C-C) Duplex

In the course of studying BLM-DNA interactions, we found that d(G-G-G-G-A-G-C-T-C-C-C-C) was selectively cleaved at the G4pA5 site rather than the G6pC7 site, which is usually a preferred site for BLM. We estimated the cleavage efficiency of G4pA5 of d(G-G-G-G-A-G-C-T-C-C-C-C) relative to G6pC7 of d(C-C-C-C-A-G-C-T-G-G-G-G). The GpA cleavage was 4 times as fast as the GpC cleavage in the d(G-G-G-G-A-G-C-T-C-C-C-C) duplex. The G4pA5 site was cleaved faster than the G6pC7 site of d(C-C-C-C-A-G-C-T-G-G-G-G) where specific cleavage occurred, while the cleavage at the G6pC7 site of d(G-G-G-G-A-G-C-T-C-C-C-C) was suppressed. These results clearly show that the sequence preference of BLMs, GpT or GpC > GpA , was reversed for d(G-G-G-G-A-G-C-T-C-C-C-C). The G-G-G-G-A-G-C-T-C-C-C-C sequence is very unique: the 5′-half contains six consecutive purines and the 3′-half contains six consecutive pyrimidines. The sequence can be derived from d(G-G-G-G-G-G-C-C-C-C-C-C) with one G : C-to-A : T base pair replacement. We assume that such altered specificity of BLM cleavage is due to the unique conformation of the d(G-G-G-G-A-G-C-T-C-C-C-C) duplex itself since BLM does not perturb the conformation of the DNA duplex upon complex formation.

To clarify the correlation between the structure of DNA and the cleavage efficiency by BLM, we determined the fine structure of d(G-G-G-G-A-G-C-T-C-C-C-C) by a combination of 2D NMR data and restrained molecular dynamics calculation (*27*, *28*). In addition, we attempted to explain the altered sequence specificity of BLM cleavage taking the calculated structure and the minor-groove binding model into consideration.

2. Unusual Sugar Puckering in d(G-G-G-G-A-G-C-T-C-C-C-C) Duplex

Using standard methods, all imino proton and most cytosine amino proton resonances were assigned. Non-exchangeable protons were assigned from NOESY spectra using the standard sequential assignment method (*29*), and the assignments were confirmed by DQF-COSY spectra. From analysis of the DQF-COSY spectrum (Fig. 7), we found unusual cross peaks between H2″ and H3′ with intermediate intensities (boxed in the figure). These peaks are of C7H2″-H3′ and C9H2″-H3′.

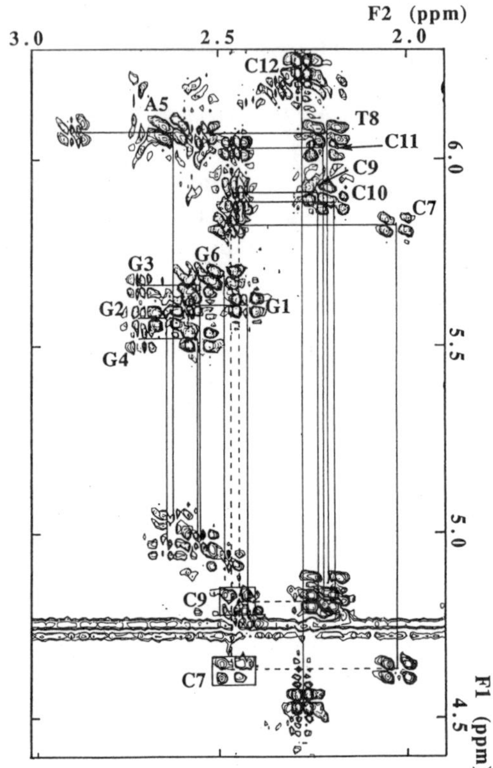

Fig. 7. Expansion of the DQF-COSY spectrum of d(G-G-G-G-A-G-C-T-C-C-C-C) at 30°C, pH 6.8. The H1′-H2′ and H1′-H2″ cross peaks are connected by horizontal lines. The H1′-H2′ and H3′-H2′ cross peaks are also connected by vertical lines. The numbering of each residue is indicated. For C7 and C9 spin-systems, H3′-H2″ cross peaks are boxed and connected by dashed lines.

Such correlation is not observed in a usual B-DNA structure, in which deoxyriboses adopt C1′-exo puckering for pyrimidine, and C2′-endo for purine residues (*30*, *31*). The C7 and the C9 residues in d(G-G-G-G-A-G-C-T-C-C-C-C) are assumed to adopt an O4′-endo puckering. The simulation study of the COSY cross peak pattern (*32*) was also consistent with such result.

3. Unique Structure of the Polypurine:Polypyrimidine Junction Affects BLM-DNA Interaction

The calculated structures obtained from B- and A-form initial structures are shown in Fig. 8.

The minor-groove view of the G6pC7 site in the standard B-DNA structure ($\langle b_{ini} \rangle$) and the calculated structure ($\langle b_{rmd}' \rangle$) are compared in Fig. 9a and 9e, respectively. The minor-groove of the G6pC7 site of d(G-G-G-G-A-G-C-T-C-C-C-C) was substantially widened and distorted when compared with that of the standard B-DNA. We analyzed the base stacking pattern at each base-pair step in the calculated structure. In comparison with the standard B structure (Fig. 9b and 9c), the base overlap between purine bases, between G4 and A5, and between A5 and G6, is improved (Fig. 9f and 9g, respectively), and that between pyrimidine bases is reduced. These phenomena are also observed in the GpG steps of this duplex (data not shown). Stabilization by continuous base stacking between these purine residues may be a driving force in forming the unique conformation of the d(G-G-G-G-A-G-C-T-C-C-C-C) duplexes. In contrast, step G6pC7, which is the only purine-pyrimidine step in the duplex, shows very small base overlapping (Fig. 9h).

In our minor-groove binding model of BLM, the DNA binding region of BLM, the bithiazole and the terminal cationic residue covers approximately 4 base pairs (Fig. 6). We speculated on the recognition mechanism of BLM using two hydrogen bonds, one with G-C2NH_2 and the other with pyrimidine C2O. We compared locations of these atoms in the standard B-DNA and the calculated structure (Fig. 9d and 9h, respectively). In the standard B-DNA structure, in which the GpC site is efficiently cleaved, the C7O2 atom is located just below the G6N2 atom. In the $\langle b_{rmd}' \rangle$ structure, the C7O2 atom is located at the outer side of the minor groove and not below the G6N2 atom. The distances between

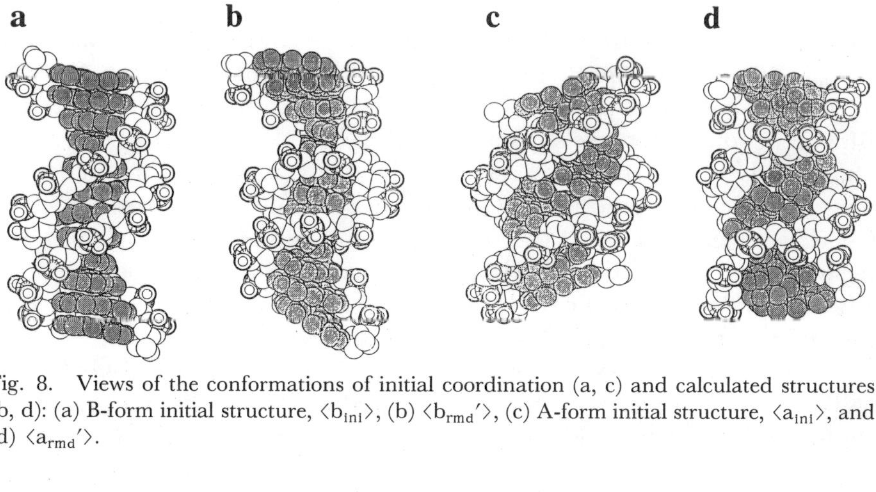

Fig. 8. Views of the conformations of initial coordination (a, c) and calculated structures (b, d): (a) B-form initial structure, $\langle b_{ini} \rangle$, (b) $\langle b_{rmd}' \rangle$, (c) A-form initial structure, $\langle a_{ini} \rangle$, and (d) $\langle a_{rmd}' \rangle$.

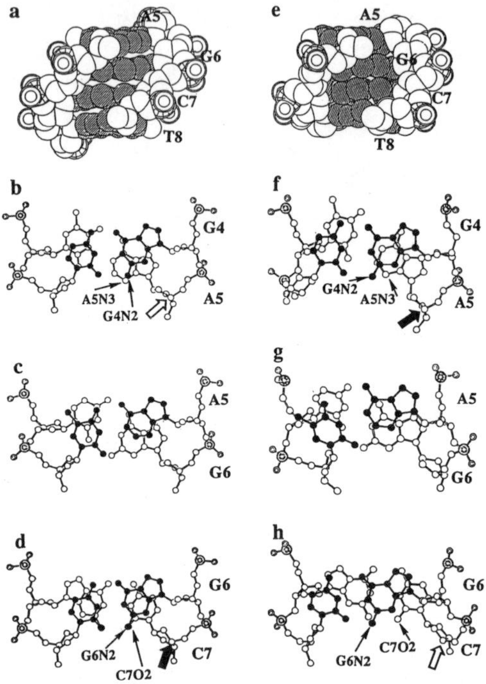

Fig. 9. Comparison of the minor-groove views and the base stacking around the cleavage sites of ⟨b_{ini}⟩ and ⟨b_{rmd}'⟩ structures. a–d: for the standard B-DNA structure (⟨b_{ini}⟩); e–h: for the calculated structure (⟨b_{rmd}'⟩). a and e: minor-groove view of A5pG6pC7pT8: A5pG6pC7pT8 site; b and f: G4pA5:T8pC9 step; c and g: A5pG6:C7pT8 step; d and h: G6pC7:G6pC7 step.

G6N2H and C7O2 are about the same, 3.10 Å and 2.93 Å in ⟨b_{ini}⟩ and ⟨b_{rmd}'⟩ structure, respectively. However, the distances between G6N2H and C7H4′, which is primarily attacked by free radicals produced at the BLM reaction center, are substantially different, 3.60 Å and 5.26 Å in the ⟨b_{ini}⟩ and ⟨b_{rmd}'⟩ structure, respectively. Thus, it is probable that BLM does not efficiently bind to the G6pC7 site of the d(G-G-G-G-A-G-C-T-C-C-C-C) duplex, or even if the drug can bind to the site, BLM may not mediate cleavage at the C7H4′ position, because the reaction center of BLM is not located at the 'right' position with an appropriate distance or direction to attack C7H4′.

We also tried to explain why the G4pA5 site in the d(G-G-G-G-A-G-C-T-C-C-C-C) duplex was efficiently cleaved by BLM, whereas that of the standard B-DNA was not. In this case, we focused on the N3 atom of adenine base. A-N3 is known as a good proton acceptor facing the minor groove, especially in the cases of netropsin or distamycin, which are well-studied minor-groove binders. In Fig. 9f, the A5N3 atom is located below the G4N2 atom, and in the standard B-DNA, the A5N3 atom is also located just below this atom (Fig. 9b). We assume that the difference in the direction of hydrogen bond formation between the N3 atom of adenine and the O2 atom of pyrimidine is critical for the base recognition.

SUMMARY

We have studied mismatch base-pairing structures involving dI in synthetic DNA oligonucleotides by NMR. It was found that a hypoxanthine base can form hydrogen-bonded base pairs with all four bases, C, A, T, and G, in the duplexes. For dI:dA pairing, I(anti):A(syn) and I(anti):A(anti) conformations coexist in the d(C-G-C-I-A-A-T-T-A-G-C-G) duplex, whereas only I(anti):A(anti) conformer is found in the d(G-G-I-A-C-C) duplex. During the course of studying BLM-DNA interactions, we found that d(C-C-C-C-A-G-C-T-G-G-G-G) was specifically cleaved at the G6pC7 site but that d(G-G-G-G-A-G-C-T-C-C-C-C) was selectively cleaved at the G4pA site rather than the G6pC7 site. The three dimensional structure of the d(G-G-G-G-A-G-C-T-C-C-C-C) duplex was determined using a combination of 2D-NMR techniques and restrained molecular dynamics calculation. The results show a distorted B-DNA structure with a widened minor groove at the G6pC7 site. These findings suggest that local environmental differences in the base sequences affect the mismatch base-pair conformation equilibrium and specificity in the drug-DNA interactions.

Acknowledgments
We thank Dr. Tomohisa Takita for providing the BLM-B2 sample. This work was supported in part by a Grant-in-Aid for Specially Promoted Research from the Ministry of Education, Science, and Culture of Japan.

REFERENCES

1. E. Ohtsuka, S. Matsuki, M. Ikehara, Y. Takahashi, and K. Matsubara: *J. Biol. Chem.* **260**, 2605–2608 (1985).
2. Y. Takahashi, K. Kata, Y. Hayashizaki, T. Wakabayashi, E. Ohtsuka, S. Matsuki, M. Ikehara, and K. Matsubara: *Proc. Natl. Acad. Sci. U.S.A.* **82**, 1931–1935 (1985).

3. F.H.C. Crick: *J. Mol. Biol.* **19**, 548–555 (1966).
4. S. Uesugi, Y. Oda, M. Ikehara, Y. Kawase, and E. Ohtsuka: *J. Biol. Chem.* **262**, 6965–6968 (1987).
5. Y. Oda, S. Uesugi, M. Ikehara, Y. Kawase, and E. Ohtsuka: *Nucl. Acids Res.* **19**, 5263–5267 (1991).
6. P.W.R. Corfield, W.N. Hunter, T. Brown, P. Robinson, and O. Kennard: *Nucl. Acids Res.* **15**, 7935–7949 (1987).
7. C. Carbonnaux, G.V. Fazakerley, and L. Sowers: *Nucl. Acids Res.* **18**, 4075–4081 (1990).
8. W.B.T. Cruse, J. Aymani, O. Kennard, T. Brown, A.G. C. Jack, and G.A. Leonard: *Nucl. Acids Res.* **17**, 55–72 (1989).
9. F.H. Martin, M.M. Castro, F. Aboul-ela, and I. Tinoco, Jr.: *Nucl. Acids Res.* **13**, 8927–8938 (1985).
10. Y. Kawase, S. Iwai, H. Inoue, K. Miura, and E. Ohtsuka: *Nucl. Acids Res.* **14**, 7727–7736 (1986).
11. Y. Kawase, S. Iwai, and E. Ohtsuka: *Chem. Pharm. Bull.* **37**, 599–601 (1989).
12. D.J. Patel, S.A. Kozlowski, S. Ikuta, and K. Itakura: *Biochemistry* **23**, 3207–3217 (1984).
13. T. Brown, W.N. Hunter, G. Kneale, and O. Kennard: *Proc. Natl. Acad. Sci. U.S.A.* **83**, 2402–2406 (1986).
14. L.-S. Kan, S. Chandrasegaran, S.M. Pulford, and P. Miller: *Proc. Natl. Acad. Sci. U.S.A.* **80**, 4263–4265 (1983).
15. G.G. Prive, U. Heinemann, S. Chandrasegaran, L.-S. Kan, M.L. Kopka, and R.E. Dickerson: *Science* **238**, 498–504 (1987).
16. H.R. Drew and R.E. Dickerson: *J. Mol. Biol.* **151**, 535–556 (1981).
17. S.M. Hecht: "Bleomycin: Chemical, Biochemical and Biological Aspects," Springer-Verlag, New York (1979).
18. J. Stubbe and J.W. Kozarich: *Chem. Rev.* **87**, 1107– 1136 (1987).
19. T.E. Booth, T.T. Sakai, and J.D. Glickson: *Biochemistry* **22**, 4211–4217 (1983).
20. L.F. Povirk, M. Hogan, and N. Dattagupta: *Biochemistry* **18**, 96–101 (1979).
21. M.J. Levy and S.M. Hecht: *Biochemistry* **27**, 2647–2650 (1988).
22. R.E. Dickerson: "Mechanism of DNA Damage and Repair: Implications for Carcinogenesis and Risk Assessment in Basic Life Sciences" (Ed. by M.G. Smi and L. Grossman) Vol. 38, pp. 245–255, Plenum, New York (1986).
23. J. Kuwahara and Y. Sugiura: *Proc. Acad. Natl. Sci. U.S.A.* **85**, 2459–2463 (1988).
24. H. Hiroaki, T. Nakayama, M. Ikehara, and S. Uesugi: *Chem. Pharm. Bull.* **39**, 2780–2786 (1991).
25. H. Hiroaki, T. Nakayama, M. Ikehara, and S. Uesugi: *J. Biomol. Struct. Dyn.* **8**, a085 (1991).
26. H. Hiroaki, S. Uesugi, and S. Fujii: *Nucl. Acids Symp. Ser.* **25**, 147–148 (1991).
27. G.M. Clore, H. Oschkinat, L.W. McLaughlin, F. Benseler, C.S. Happ, E. Happ, and A.M. Gronenborn: *Biochemistry* **27**, 4185–4197 (1988).
28. M. Katahira, H. Sugeta, Y. Kyogoku, and S. Fujii: *Biochemistry* **29**, 7214–7222 (1990).
29. K. Wüthrich: "NMR of Proteins and Nucleic Acids," John Wiley & Sons, New York (1986).
30. K.V. R. Chary, R.V. Hosur, G. Govil, T. Zu-kun, and H.T. Miles: *Biochemistry* **26**, 1315–1322 (1987).
31. K.V.R. Chary, R.V. Hosur, G. Govil, C.Q. Chen, and H.T. Miles: *Biochemistry* **27**, 3858–3867 (1988).
32. M. Gochin, G. Zon, and T.L. James: *Biochemistry* **29**, 11161–11171 (1990).

Molecular Structure and Life (Ed. by Y. Kyogoku and Y. Nishimura), pp. 43–55, Japan Sci. Soc. Press, Tokyo/CRC Press, Boca Raton (1992)

Curvature and Flexibility of DNA: Sequence-directed Effects Seen from a Structural Mechanics Viewpoint

C.R. CALLADINE*[1] and H.R. DREW*[2]

Department of Engineering, University of Cambridge[1]
CSIRO Division of Biomolecular Engineering, Laboratory for Molecular Biology[2]

This paper is mainly about the structure of double-helical DNA, and how certain "gross" features of the shape of a DNA molecule depend on the sequence of bases within it. These "gross" features include the intrinsic curvature of the molecule, the flexibility or ease of bending the molecule under stress, and items such as the widths of the major and minor grooves between the helical sugar-phosphate chains.

I. THE BASIC DOUBLE-HELICAL STRUCTURE OF DNA

The primary structure of DNA involves a sugar-phosphate chain, with side groups consisting of the four bases A, G, C, and T. Two of these chains connect along their length by cross-chain hydrogen bonds between the bases to form a simple ladder; and the sequence of bases on the two chains is usually self-complementary, thus satisfying the rule "A pairs with T and G pairs with C" (*1*). All four types of base are hydrophobic, while the longitudinal sugar-phosphate chains are hydrophilic.

*[1] Cambridge CB2 1PZ, U.K.
*[2] Sydney NSW 2113, Australia

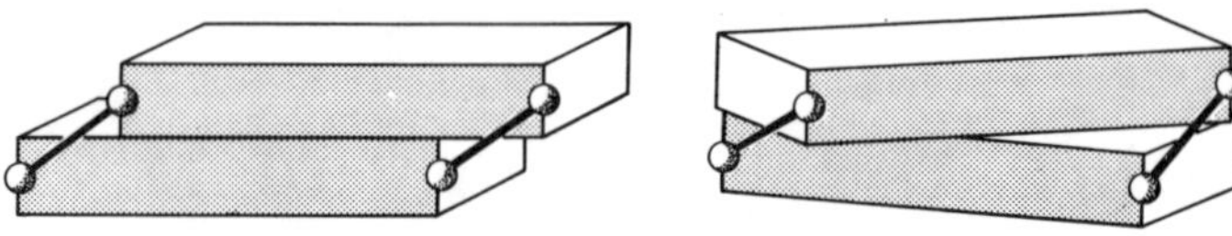

(a) (b)

Fig. 1. Two possible arrangements for closing the gap between base pairs, including (b) helical Twist.

It is not difficult to see why the paired chains of a DNA ladder should prefer to assemble into a straightforward, regular, double-helical structure. The phosphate-to-phosphate distance along the chains is around 6 Å as a consequence of the conformation of the covalently-bonded chain structure. But the van der Waals thickness of the flat bases is only about 3.3 Å. Thus in a "straight ladder" configuration there would be *holes* between successive base pairs, filled by solvent molecules. But the flat surfaces of all of the bases are hydrophobic; and so the bases will try to stack onto each other in direct contact, thereby excluding the solvent. There are two ways of doing this, as shown in Fig. 1 for two adjacent base pairs. The first way forms a plane "skew-ladder". The second way forms a helix, and this is evidently more satisfactory. Using the dimensions given above, and taking the effective width of the ladder between phosphates as 18 Å, we find by simple geometry an angle of helical twist per step of 32° (*2*), which is close to what is observed in practice for most DNA double helices, whether right-handed or left-handed.

Apart from a few exceptions, most of the DNA structures which have been determined by X-ray crystallography show a basic right-handed double-helical structure (*3*, *4*). The many right-handed double helices seen in these crystals are roughly uniform in structure on a superficial level, but there are striking local variations. For example, the angle of helical twist at a dinucleotide step can be anywhere within a range from 28° to 40°, with an average of around 34°.

II. THE OVERALL SHAPE OF THE DNA FOLLOWS FROM THE LOCAL CONFIGURATION OF BASE-PAIR STEPS

Let us examine the geometrical relationship between the configuration of the dinucleotide steps on the one hand and the overall shape of the double-helical molecule on the other. We begin by removing the sugar-phosphate chains entirely, and replacing the base pairs by plain, rectangular blocks; and then we consider the molecule as a series of *steps* built one by one on top of each other. To preserve our sense of orientation, we always darken the minor-groove edges of the rectangular blocks.

We can now describe the geometry of a given dinucleotide step by means of the six *degrees of freedom* which one rigid object has with respect to another in three-dimensional Euclidean space: three are translations and three are rotations (*5*). It turns out that the van der Waals stacking effects and the presence of the sugar-phosphate chains in real DNA make three of these degrees of freedom practically inoperative. The three remaining and significant degrees of freedom are shown schematically in Fig. 2 as Roll and Slide, together with Twist as already seen in Fig. 1. Roll is a hinge-like relative rotation about the mean long axis of the base pairs; and its positive sense opens up the minor-groove side of the base pairs. Slide* is a relative translation along the same long axis.

Figure 3 shows an array of four pictures of a complete helical turn of DNA, built from 10 identical steps having Twist$\approx$36°. In (a) we have Roll=0° and Slide=0 Å. All of the base-pair blocks lie in parallel planes, perpendicular to the line of their centres. Picture (b) shows the consequence of keeping Roll=0° but moving to Slide=2 Å: the assembly now becomes wider, and in an end-on or top view we would see a hole down the middle of the stack. In (c) we show Slide=0 Å but Roll=12°: and now the base-pair blocks tilt with respect to the axis of the molecule. Finally, in (d) we see the effect of adding Slide to Roll. Once again the width of the stack increases due to the addition of Slide, just as it did in the case of helix (b); but here, in addition, Slide produces an overall *shortening*, because the base-pair blocks are already tilted with respect to the axis.

The four pictures show portions of four different versions of *straight*, uncurved DNA; and the values of Roll and Slide have been chosen in such a way that models (a) and (d) correspond, respectively, to the

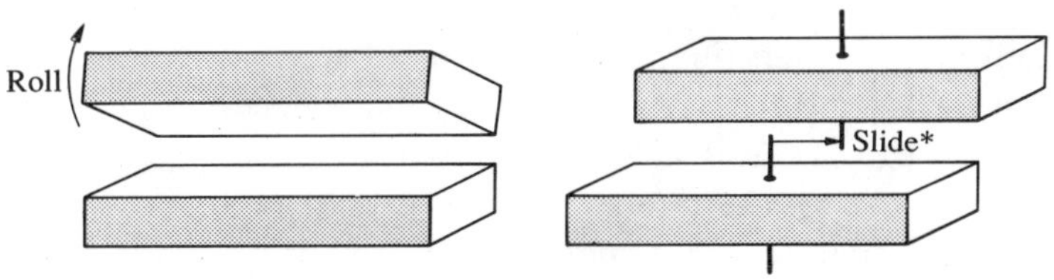

Fig. 2. Roll and Slide in a base-pair step.

* The variables which we have used here correspond to internationally agreed symbols (*9*) as follows: Roll=ρ, Slide=$-D_y$, Twist=ω.

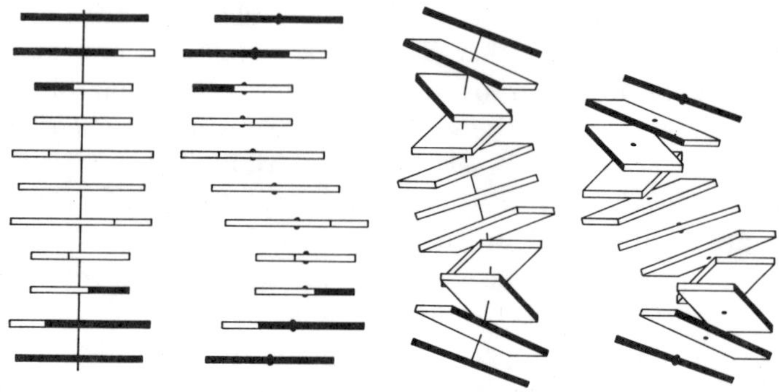

Fig. 3. Effects of uniform Roll (R) and Slide (S*) on a helical turn of DNA.

well-known "B" and "A" forms of DNA seen in fibres by X-ray diffraction. Various well-known "external" features of the transition from "B" to "A"—*viz.* the overall shortening, the tilt of the base pairs relative to the axis, and the overall increase in diameter—are produced perfectly by our "interior" changes of Roll and Slide. The only feature not modelled here is the change of helical repeat from about 10 to about 11 steps per helical turn; but this is straightforward to fix.

Note that, if we were to put back the sugar-phosphate chains, we should see that the "A" form has a wider minor groove than the "B" form.

III. CURVED DNA

Now, if we wish to have a curved piece of DNA, we must alter the Roll angles at the dinucleotide steps in a *non-uniform* way along the length of the molecule. Let us keep the helical Twist at $\approx 36°$ in each step, for the sake of simplicity. Then the insertion of a substantial Roll angle at every tenth step will produce a plane—if somewhat "angular"—curve, as shown in Fig. 4 (a). The Roll angles shown here have been set at 45°, so that the curvature works out at 45° per helical turn, or 4.5° per dinucleotide step on average, as when DNA is wrapped around the histone octamer of chromatin to make a "nucleosome core" (*6*). In fact, this is not such an unrealistic model of curvature, because isolated 45° kinks have been seen in the complex of DNA with the CAP protein from *Escherichia coli* (*7*).

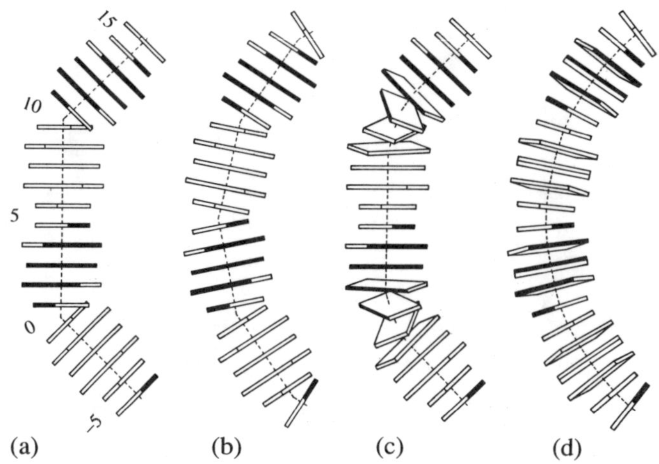

Fig. 4. Various Roll patterns which produce curvature in DNA.

There are several different ways of disposing the sequence of Roll angles in order to achieve the same overall level of curvature of 45° per helix turn, and some of these are also shown in Fig. 4. Thus, in scheme (b) there are two steps in each helical repeat with Roll of 22.5° instead of one at 45°; but in this case they are alternately + and − because the orientation of the minor groove changes by 180° every five steps. In scheme (c) Roll angles of 14° and 0° have been imparted to successive groups of five steps. This arrangement is just like taking five steps from Fig. 3 (d) and (a) in turn: and one can see how the progressive change in helical direction comes about. Another scheme for imparting curvature is shown in (d). Here, the Roll angles vary in the fashion of a cosine-wave along the length of the molecule. If we make the period of the cosine-wave match the helical repeat of the DNA, we obtain a smooth plane curve; and then Roll need vary by only ±9°, as shown, to give again a curvature of 4.5° per step.

But the varieties shown here do not exhaust all possible schemes for imparting curvature to DNA. For instance, the overall curvature would not be changed in any of these examples if a *constant* Roll of arbitrary magnitude were added at every step.

Figure 5 shows another representation of the four configurations of Fig. 4, as a schematic plot of Roll angle against step number. What is common to all of these plots is that the *Fourier component* of Roll at a period equal to the helical repeat (here, *10*) has the same value through-

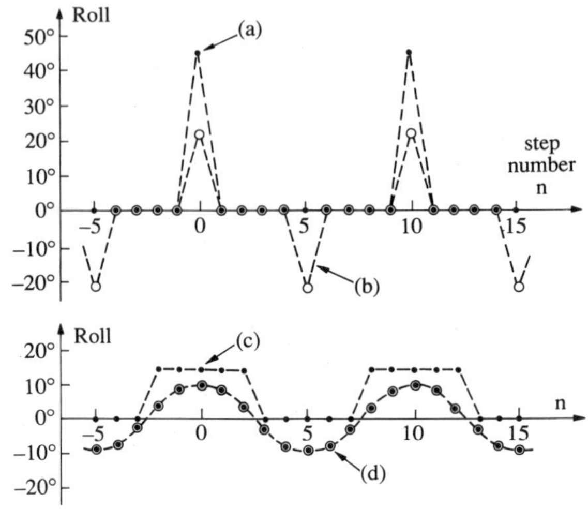

Fig. 5. Plots of Roll against step number: *cf* Fig. 4.

out. It is precisely this Fourier component which produces the overall curvature (*8*).

IV. PROPELLER TWIST

Thus far we have idealised the base pairs as simple, rectangular blocks. In order to make more progress in understanding the structure of DNA, we need to introduce another geometrical parameter known as *Propeller twist.* It turns out that in real DNA the base pairs are not truly planar: instead, they include, typically, 10° of *Propeller twist.* This is shown schematically in Fig. 6: here, two base pairs, stacked with zero helical Twist, have Propeller twist of the sense which is usually observed in DNA oligomers (but which is now regarded as negative (*9*)).

What is the source or origin of Propeller twist in DNA? As yet, there is no single, simple, and totally convincing answer to this question; but simple 3-D geometrical models suggest that water has less access, on the whole, to the hydrophobic surfaces of the bases when the base pairs are Propeller-twisted (*2*).

When Propeller-twisted base pairs engage in Slide (see Fig. 2), the twisted shape produces a screw-like motion, so that Slide also produces Roll. Indeed, a broad correlation between Slide and Roll has been observed in the dinucleotide steps of crystallised DNA oligomers (*5*).

So far, we have drawn the separate bases in a pair as being of equal

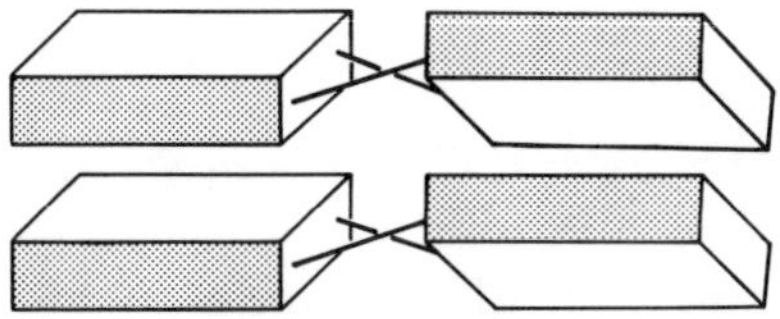

Fig. 6. Propeller-twist in base pairs.

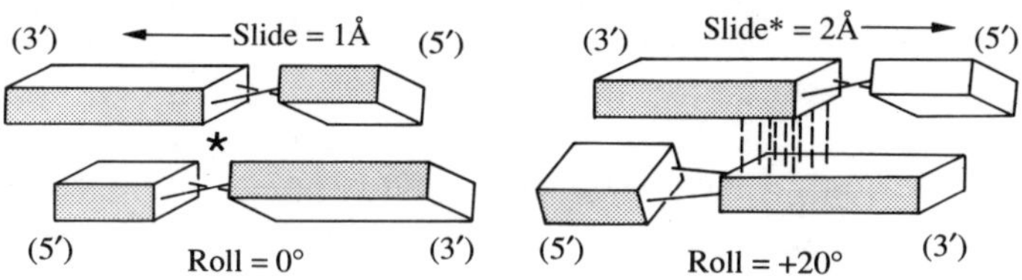

Fig. 7. Two distinct stacking arrangements for pyrimidine-purine steps.

size. But the purines A and G are larger than the pyrimidines T and C. This has a steric consequence which is shown schematically in Fig. 7. In a pyrimidine-purine step (CG, CA, TG, or TA) with Propeller twist, there tends to be a "cross-chain clash" between the large purines on the minor-groove side of the pairs, which holds the step in a low-slide configuration as in (a) (*10*). Alternatively, as shown in (b), the purines can ride over each other, so that the step adopts a high Slide, high Roll configuration (*5*). Pyrimidine-purine steps are generally found in crystallised oligomers to be either in one configuration or the other, but not in intermediate positions: they show *bistable* behaviour, of a kind not displayed by other steps. The origin of bistability of pyrimidine-purine steps lies partly in van der Waals, and partly in electrostatic effects (*11*), which we shall not explore here; but in any case the sketches of Fig. 7 provide a good mnemonic for the permissible manoeuvres.

Propeller twist tends to be somewhat hindered, or damped down, by the cross-chain interaction of bases as in both parts of Fig. 7. Where the hydrogen-bonded moieties of neighbouring base pairs are in good alignment, as shown schematically in Fig. 6, the Propeller twisting is less restrained. Propeller twist has a particularly high value—around (−)20°—in the sequence AAAAAA (*12*). This is thought to be due in part to the formation of an additional cross-chain hydrogen bond on the major-groove side between adjoining A-T base pairs; and it is known from other studies that a long run of A or of T on one strand of a DNA molecule confers unusual resistance to bending about proteins.

V. SEQUENCE-SPECIFIC DINUCLEOTIDE STACKING PATTERNS

Let us now consider the general question of the relationship between the *sequence* of a dinucleotide step and the *configuration* of the step.

There are altogether ten distinct dinucleotide sequences AA, AG, AT, AC, *etc.* (not 16, since TT is physically the same as AA, *etc.*), and it would be simple and convenient if each were associated with a unique set of Roll, Slide, and Twist values. But reality is not as simple as this. We already know that pyrimidine-purine steps have ambivalent, or bistable, configurations. Furthermore, it is known that the configuration of some steps can be affected by the sequence and configuration of neighbouring steps, either through the transmission of Slide along the sugar-phosphate chains, or through the Propeller twist in a base pair, which is shared by neighbouring base pairs.

How can we obtain basic data about the relationship between the base sequence and the local configuration? So far, four different sources of such information have been found by scientists, as follows. First, we can examine the configuration at specific dinucleotide steps in oligomers by single-crystal X-ray methods. No very clear-cut sequence/structure patterns have emerged from this kind of approach, however. The only firm result so far from the X-ray work seems to be that high-Roll steps tend to have high Slide and low Twist, and *vice versa* (*4*).

A second source of information comes from the sequences of DNA molecules, about 150 base pairs long, which have been wrapped around histone octamers in nucleosomes to a curvature of about 45° per helical turn (*13*). It might be supposed that dinucleotide steps which can easily adopt a high Roll or low Roll configuration will tend, on average, to prefer locations where the minor-groove of the DNA faces outwards or inwards, respectively (see Fig. 4). A detailed study of about 200 different samples of chicken nucleosome core DNA has shown that some particular dinucleotide sequences—such as AA or TT—have a strong preference for low Roll, while others—such as GC—have a strong preference for high Roll. Steps TG/CA seem to prefer *either* high- *or* low-Roll locations, rather than intermediate positions. And certain other dinucleotide sequences show little preference for any position. In general, specific sequences of DNA can confer certain kinds of flexibility on a double helix: often the DNA can bend only in one direction, like a finger.

A third source of information about the relationship between sequence and step configuration comes from the anomalous running of particular repeating-sequence DNA molecules through electrophoretic gels (*14*, *15*). The raw data are not easy to interpret, but it is generally agreed that the dinucleotide steps AA and TT, especially in runs, have unusually low Roll values.

The fourth source of information about base-stacking arrangements comes from theoretical chemical studies of stacking interactions at particular kinds of base-pair steps, in relation to a variety of possible local configurations of Roll, Slide, and Twist, *etc.* The various energetic factors which are involved here have been clarified recently (*11*), and we expect soon to see some quantitative results coming from this approach.

It is by no means a straightforward matter to fit together the data which flow from these four different sources. We need some sort of conceptual framework by means of which the different lines of evidence may be interpreted, and we believe that *structural mechanics* provides such a scheme. This enables us, for example, to distinguish between the slight intrinsic curvature of relaxed DNA as measured in gel-running assays, and the much greater curvature of DNA that is subjected to external force, such as when it wraps around a histone octamer; this is an important distinction (*16*).

The most precise information about the geometry of base-stacking arrangements comes from single-crystal X-ray assays. But are DNA oligomers in general in a *relaxed* or a *stressed* state when they are fitted together into a crystal? It is difficult to answer this question. The fact that the range of Roll angles observed in crystallised oligomers ($\pm 10°$, typically) is about the same as that needed for flexure around a nucleosome core might suggest that oligomers in crystals are somewhat stressed; but it may be, alternatively, that much of this range would be found in relaxed structures also.

VI. THE BINDING OF DNA TO 434 REPRESSOR

The X-ray structure of a complex between the 434 repressor protein and a 14 base-pair DNA double helix of sequence ACAATATATATTGT shows that the protein contacts the bases within the major grooves of the DNA at the two ends of the 14-mer, but that it does not contact any bases in the central region (*17*, *18*). This is shown schematically in Fig. 8. Nevertheless, the sequence of bases in this central region contributes to the specificity of the interaction. Thus, replacement of the central ATAT by GTAT, GTAC, or ACGT weakens the binding affinity by a factor of 2.5, 5, or 50, respectively. How is this accomplished?

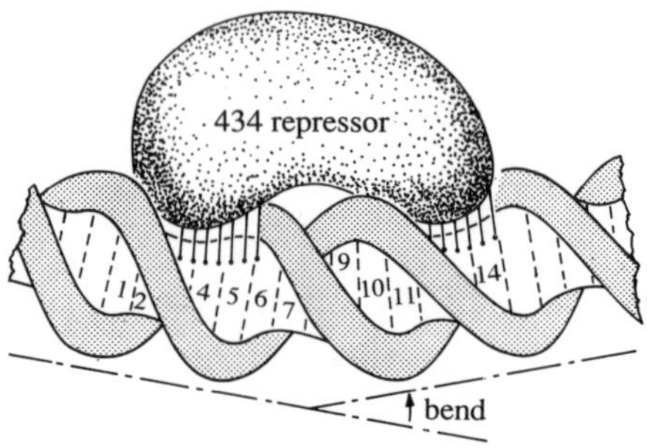

Fig. 8. Binding of DNA to 434 repressor.

The DNA curves substantially about the protein, such that the minor groove in the central region ATAT lies along the inside of the curve. The statistics of dinucleotide step preferences in bending around histone octamers, as mentioned above, may be used to predict the ease of bending the central portion of the oligomer according to its sequence; and there is a good correlation between the predicted values and the observed binding of different sequences to the 434 repressor (*19*).

VII. THE DOCK-AND-PROBE CONCEPT OF RECOGNITION

The example in VI illustrates what we believe to be a general two-stage process in the recognition of a DNA sequence by a protein. First, the protein recognises some *gross* feature of the DNA, in order to *dock* successfully onto the DNA. Then, having once docked, the protein can *probe* the local surface of the DNA in the contact-regions in order to seek a large number of hydrogen-bonded connections between the protein and the DNA—the kind of contacts which are seen clearly in high-resolution X-ray studies of co-crystals of proteins bound to DNA oligomers (*20*, *21*). Only if both conditions are satisfied will the binding be strong. These two steps might not, of course, necessarily happen in the order which the above description suggests: all contact and recognition processes occur against a background of brownian buffeting. Nevertheless, we think that the two stage "dock-and-probe" pattern may well be a general one, even though the details of the "docking" and "probing" arrangements differ greatly among various examples.

In this connection we now mention some other gross features of DNA which are a consequence of base-stacking arrangements, and we suggest examples of their use for purposes of docking with a protein.

TABLE I
Influence of the Base-step Parameters on the Outward Appearance of DNA

Positive motion at the level of a base-pair step	Influence on the outward appearance of sugar-phosphate chains	Examples of proteins that recognise the variation in helix shape
Roll	open minor groove, close major groove	histone octamer (*13*), 434 repressor (*18*), catabolite activator (CAP) (*7*)
Slide	increase radius of helix, turn phosphates into major groove	zinc-finger proteins such as TFIIIA (*22*) or Zif268 (*23*)
Twist	increase contour length of sugar-phosphate chains; need fewer base pairs to make one helix turn	various polymerases and ethidium bromide recognise a *decrease* in twist
Propeller twist	close minor groove, open major groove	netropsin, distamycin, glucocorticoid-receptor (*24*)

In II we described the way in which the addition of uniform *Roll* to the dinucleotide steps in DNA alters the disposition of the sugar-phosphate chains in DNA, thereby *opening* the minor groove. It is not difficult to see that a uniform increase of *Propeller twist* has a broadly similar effect, but in the opposite sense of *closing* the minor groove. Table I lists the qualitative changes of outward appearance in the sugar-phosphate chains which are a consequence of various simple manoeuvres at the base-step level: these involve all four of the main parameters Roll, Slide, Twist, and Propeller twist. In each case the effect on the outward appearance of the chains is straightforward to understand by means of simple physical models. In the right-hand column we tentatively identify proteins which recognise such changes in gross helix shape in the "docking" phase of recognition.

In some respects Table I is crude: for example, the part played by increased Roll in relation to the binding of DNA to a histone octamer is connected with changes of *curvature* rather than changes in the groove widths. Nevertheless, the table as a whole lends support to the idea that the "docking" of proteins with DNA involves some sequence-dependent "gross" features of the molecule.

SUMMARY

Certain "gross" features of the DNA molecule depend on the sequence of bases within it. These include (i) intrinsic curvature of relaxed DNA, (ii) flexibility under stress, and (iii) width of the grooves between the sugar-phosphate chains.

We explain how the base-step stacking geometry determines the external form of the molecule, *e.g.* whether the DNA takes the "B" or "A" form, or whether it is straight or curved. We discuss the way in which the sequence of a dinucleotide step can determine the geometry of the step, and we argue that considerations of *mechanics* are needed in order to understand and interpret evidence on these matters which comes variously from single-crystal X-ray structures, from nucleosome core DNA sequences, from gel electrophoresis of repeated-sequence DNA, and from theoretical chemistry.

Gross features play a part in the recognition of DNA sequences by proteins. We argue that the process of recognition in general involves two steps, which we call "docking" and "probing". Docking involves the recognition by the protein of a "gross" sequence-dependent feature; while probing involves a detailed examination of the electric charges on the exposed edges of the DNA bases by the "reading head" or amino acids of the protein. Thus the process of recognition involves the sequence-dependent structure of DNA at two different levels—a situation which provides for considerable subtlety in operation.

Acknowledgments
We are grateful to Dick Dickerson, Chris Hunter, Gerald Koudelka, Ben Luisi, and Max Perutz for providing data and papers ahead of publication.

REFERENCES

1. J.D. Watson and F.H.C. Crick: *Nature* **171**, 737–738 (1953).
2. H.R. Drew and C.R. Calladine: "Understanding DNA: the Molecule and How It Works," Academic Press, in press.
3. H.R. Drew, M.J. McCall, and C.R. Calladine: *Annu. Rev. Cell Biol.* **4**, 1–20 (1988).
4. R.E. Dickerson: DNA structure from A to Z, in press.
5. C.R. Calladine and H.R. Drew: *J. Mol. Biol.* **178**, 773–782 (1984).
6. T.J. Richmond, J.T. Finch, B. Rushton, D. Rhodes, and A. Klug: *Nature* **311**, 532–537 (1984).
7. S.C. Schultz, G.C. Shields, and T.A. Steitz: *Science* **253**, 1001–1007 (1991).
8. C.R. Calladine and H.R. Drew: *J. Mol. Biol.* **192**, 907–918 (1986).
9. R.E. Dickerson *et al.*: *EMBO J.* **8**, 1–4 (1989).
10. C.R. Calladine: *J. Mol. Biol.* **161**, 343–352 (1982).
11. C.A. Hunter and J.K.M. Sanders: *J. Am. Chem. Soc.* **112**, 5525–5534 (1991).

12. H.C.M. Nelson, J.T. Finch, B.A. Luisi, and A. Klug: *Nature* **330**, 221–226 (1987).
13. S.C. Satchwell, H.R. Drew, and A.A. Travers: *J. Mol. Biol.* **191**, 659–675 (1986).
14. C.R. Calladine, H.R. Drew, and M.J. McCall: *J. Mol. Biol.* **201**, 127–137 (1988).
15. C.R. Calladine, C.M. Collis, H.R. Drew, and M.R. Mott: *J. Mol. Biol.* **221**, 981–1005 (1991).
16. A.A. Travers: *Curr. Opin. Struct. Biol.* **1**, 114–122 (1991).
17. G.B. Koudelka, S.C. Harrison, and M. Ptashne: *Nature* **326**, 886–888 (1987).
18. G.B. Koudelka, P. Harbury, S.C. Harrison, and M. Ptashne: *Proc. Natl Acad. Sci. U.S.A.* **85**, 4633–4637 (1988).
19. H.R. Drew, M.J. McCall, and C.R. Calladine: In "DNA Topology and Its Biological Effects" (Ed. by N.R. Cozzarelli and J.C. Wang), pp. 1–56, Cold Spring Harbor Laboratory Press, Cold Spring Harbor, New York (1990).
20. M.F. Perutz: "Protein Structure: New Approaches to Disease and Therapy," W.H. Freeman & Co. Publ., New York (1992).
21. S.C. Harrison: *Nature* **353**, 715–719 (1991).
22. L. Fairall, D. Rhodes, and A. Klug: *J. Mol. Biol.* **192**, 577–591 (1986).
23. N. Paveltich and C.O. Pabo: *Science* **252**, 809–817 (1991).
24. B.F. Luisi, W.X. Xu, Z. Otwinowski, L.P. Freedman, Y.R. Yamamoto, and P.B. Sigler: *Nature* **352**, 497–505 (1991).

Molecular Structure and Life (Ed. by Y. Kyogoku and Y. Nishimura), pp. 57–66, Japan Sci. Soc. Press, Tokyo/CRC Press, Boca Raton (1992)

NMR Studies of Bent DNA, an RNA·DNA Hybrid, and a Drug-DNA Complex

Masato KATAHIRA,*[1] Ritsuko SAKAGUCHI-KATAHIRA,*[2] Hiromu SUGETA,*[2] Yoshimasa KYOGOKU,*[2] Satoshi FUJII,*[3] and Seiichi UESUGI*[3]

Faculty of Engineering, Yokohama National University,[1] Institute for Protein Research, Osaka University,*[2] Faculty of Pharmaceutical Science, Osaka University*[3]*

NMR is a powerful tool to study structures of biomolecules such as nucleic acids and proteins in solution. We present structures of bent DNA d(GGAAATTTCC)×2 (*1*), an RNA·DNA hybrid $(rA)_8 \cdot (dT)_8$ (*2*), and a drug-DNA complex chromomycin-d(GGGGCCCC)×2 (*3*) determined by NMR. A new model of DNA bending on the basis of NMR observations is proposed (*4*). The first successful application of a unique NMR method, a photo-CIDNP (chemically induced dynamic nuclear polarization) method, to the structural studies of double-stranded DNA, RNA, and a drug-DNA complex is also shown (*5*).

I. THE STRUCTURE OF BENT DNA d(GGAAATTTCC)×2

It has been demonstrated by biochemical experiments that DNAs containing oligo(dA) tracts are bent, and the biological function of DNA bending is suggested. Several different models of DNA bending have been proposed, but they are contradictory (see *6* and *7* for reviews). We

*[1] Yokohama 240, Japan
*[2] Suita, Osaka 565, Japan
*[3] Suita, Osaka 565, Japan

determined the structure of putative bent DNA d(GGAAATTTCC)×2 in solution by NMR and restrained molecular dynamics (*1*). One hundred and eighty-two interproton distance restraints derived from NOESY (two-dimensional nuclear Overhauser effect spectroscopy) experiments and torsion angle restraints for each sugar moiety derived from DQF-COSY (two-dimensional double quantum filtered correlated spectroscopy) experiments were incorporated in the calculations. Classical A- and B-DNA structures with an r.m.s.d. (root mean square deviation) of 5.5 Å between them were used as initial structures in the calculations. Final structures from the two initial structures converged to an r.m.s.d. of 0.8 Å. The structure obtained is shown in Fig. 1. It exhibits several remarkable features. The second G residue takes on the BII conformation (ε:g$^-$, ζ:t) around C3′-O3′-P bonds (*8*) rather than the standard BI conformation (ε:t, ζ:g$^-$). The sugar puckering of the residues from G1 to T8 is C1′-exo and that of C9 and C10 is O4′-endo. The combination of occurrence of the BII conformation at G2 and discontinuity of the sugar puckering between T8 and C9 causes bending at the junctions between the oligo(dA) tract and the other portions. The propeller twist of 20 degrees observed in the oligo(dA) tract is higher than in the rest of the duplex. The minor groove of the oligo(dA) tract is rather compressed, which results in a slight bending toward the minor groove within the

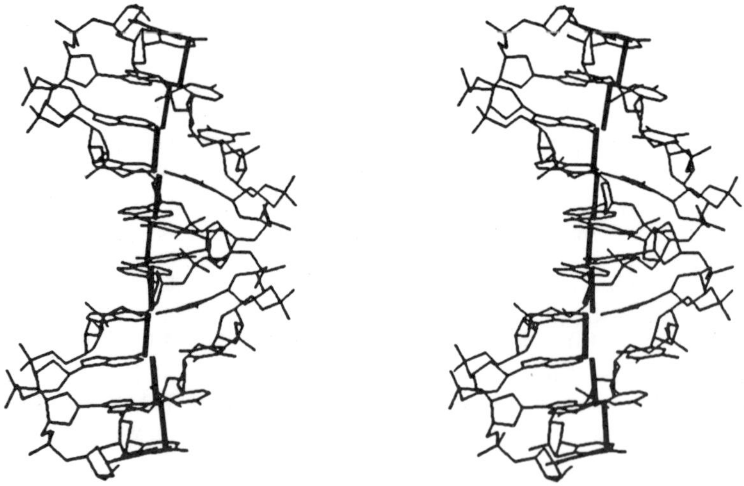

Fig. 1. Stereoview of the structure of d(GGAAATTTCC)×2. The local helical axes are indicated by heavy lines.

oligo(dA) tract. The two bendings are additive and total bending is about 20 degrees.

II. A NEW MODEL OF DNA BENDING

Further NMR studies of a series of DNAs containing the oligo(dA) tracts indicate that the minor groove of the oligo(dA) tract is compressed gradually from 5′ to 3′ along the tract (*4*, *9–11*). This is consistent with the result of cleavage by hydroxyl radicals (*12*). Additionally, it is found that the compression almost reaches a plateau level at the third adenine along the oligo(dA) tract (*4*, *11*). On the basis of these results a new model of DNA bending is proposed (*4*). The compression of the minor groove leads to bending toward the minor groove (*13–15*). Figure 2 schematically shows the direction and extent of bending for d(GAAAATTTTC)×2 and d(GTTTTAAAAC)×2. Each duplex is viewed along its helical axis. The extent of bending toward the minor groove at each residue derived from the observed compression of the minor groove is reflected by the lengths of solid line vectors. Dashed line vectors are the sums of all the vectors, which are the measures of total bending. The sum for the d(GAAAATTTTC)×2 is much greater than that for d(GTTTTAAAAC)×2, because the four longest vectors are additive in the most efficient way in the former, while the addition of those vectors tends to cancel in the latter. Thus the biochemical result that the former is bent while the latter is not (*16*) is explained by this model. In the same way, the extent of bending observed biochemically for a series of $d(A_jN_{10-j})_n \cdot d(N_{10-j}T_j)_n$ sequences (*17*) can be explained systematically. In our

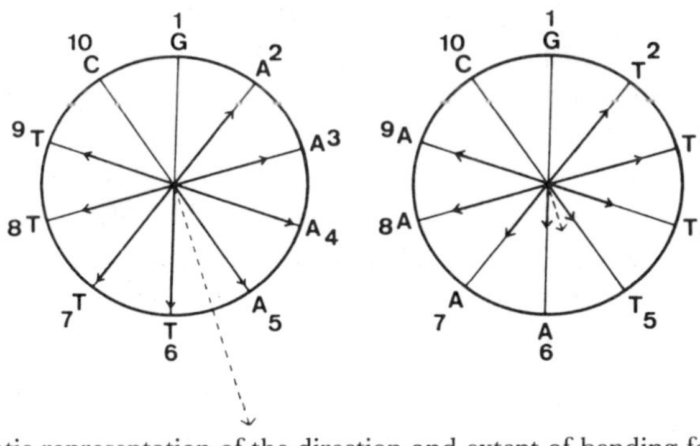

Fig. 2. Schematic representation of the direction and extent of bending for d(GAAAATTTTC)×2 (left) and d(GTTTTAAAAC)×2 (right). Solid line vectors indicate the bending toward the minor groove at each residue in the oligo(dA) tract and dashed line vectors are the sums of all the vectors.

model, the extent of bending increases gradually. The importance of difference between gradual increase of bending and uniform bending was also recognized (*4*).

As discussed in the previous section, bendings are observed at the junctions between the oligo(dA) tract and the other portions in the case of d(GGAAATTTCC)×2. Additionally, bending caused by gradual compression of the minor groove from 5′ to 3′ along the oligo(dA) tract also seems to be a key to understanding DNA bending in general.

III. THE STRUCTURE OF AN RNA·DNA HYBRID $(rA)_8 \cdot (dT)_8$

The structure of an RNA·DNA hybrid $(rA)_n \cdot (dT)_n$ has been controversial. Particularly with an rA strand, X-ray diffraction (*18*), ^{31}P NMR (*19*), and Raman spectroscopy (*20*) studies suggest that it is an A form while a ^{1}H NMR study (*21*) suggests that it is a B form. Additionally, these strands can potentially form a triple-stranded helix. We examined the structure of $(rA)_8 \cdot (dT)_8$ by ^{1}H NMR (*2*). The NOEs (nuclear Overhauser effects) observed between the H2 resonances of rA residues and the thymine imino proton resonances indicate that base pairing is of the conventional Watson-Crick type and not of the Hoogsteen type. The NOEs between the H8 of rA residues and the thymine imino proton resonances are expected for Hoogsteen base pairs. Thus the formation of the triple-stranded helix at neutral pH, where the 3rd is Hoogsteen base paired to the duplex, is excluded. Three lines of experimental evidence indicate that the rA strand takes on A form with C3′-endo sugar puckers: (1) very strong H2′($i-1$)-H8(i) NOEs, (2) relatively intense intrastrand H2(i)-H1′($i+1$) NOEs, comparable with H8(i)-H1′(i) NOEs, and (3) absence of H1′-H2′ and presence of H3′-H4′ DQF-COSY cross peaks. It was also shown that a dT strand takes on B form. Thus the structure of the RNA·DNA hybrid $(rA)_8 \cdot (dT)_8$ is very unique: the one strand takes on A form while the other strand takes B form in the duplex. The similar structure has been reported for another RNA·DNA hybrid (*22*). The presence of an intranucleotide hydrogen bonding between an H8 atom and a phosphoester O5′ atom of each adenylate residue is proposed. The structure may be stabilized by this hydrogen bonding. RNA·DNA hybrids are involved in transcription. The structure is of interest as a model for transcription.

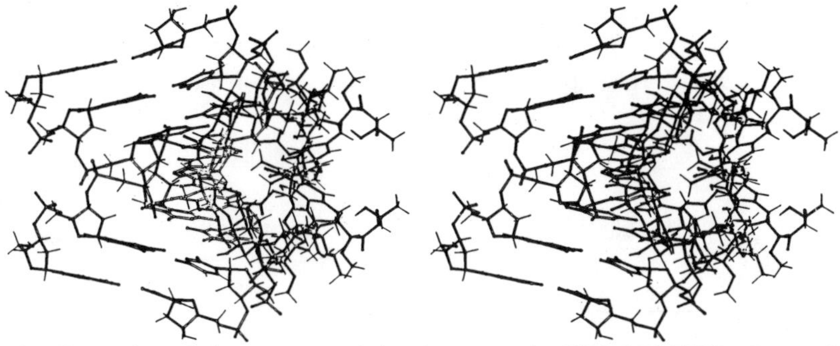

Fig. 3. Stereoview of the structure of the chromomycin-d(GGGGCCCC)×2 complex.

IV. STRUCTURE OF A CHROMOMYCIN-d(GGGGCCCC)×2 COMPLEX

Chromomycin is an antitumor drug and binds preferentially to GC-rich regions (*23*, *24*). We determined the structure of a chromomycin-d(GGGGCCCC)×2 complex on the basis of observed intermolecular NOE contacts together with intramolecular NOEs (*3*). Chromomycin binds as a symmetry-related dimer to the minor groove of the central four residues of the DNA (Fig. 3). These features are consistent with structures determined by other groups (*25*, *26*). The central four residues undergo a dramatic conformational change from B form to A form upon the binding of chromomycin. The change seems to be indispensable for accommodation of the bulky chromomycin dimer in the minor groove. Hydrogen bonding between the NH_2 group of the fourth guanine and the OH group at position 8 of chromomycin is suggested. This hydrogen bonding may explain the DNA-sequence selectivity of chromomycin.

V. APPLICATION OF A PHOTO-CIDNP METHOD TO NUCLEIC ACIDS

The photo-CIDNP method is a very unique ^{1}H NMR technique. The photo-CIDNP effect arises from the interaction of photo-excited flavin with aromatic rings of biomolecules and a following spin-sorting process acting during recombination of radical pairs. This method has been applied successfully to proteins to study their surface structure and their interaction with ligands (see *27* for a review). Its application to nucleic acids, however, has been quite limited, because CIDNP signals have not been observed for double-stranded nucleic acids (*28*, *29*). Here, we report the first observation of the CIDNP signals of double-stranded nucleic acids (*5*).

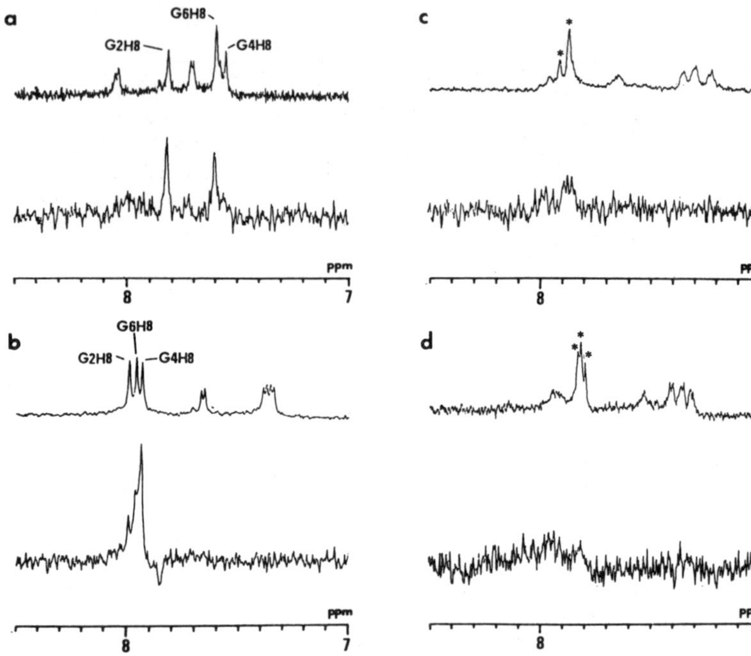

Fig. 4. One-dimensional normal (upper) and photo-CIDNP difference (lower) spectra of (a) r(CGCGCG)×2 in 0.15 M NaCl (A form), (b) d(CGCGCG)×2 in 0.15 M NaCl (B form), (c) r(CGCGCG)×2 in saturated $NaClO_4$ (Z form) and (d) d(CGCGCG)×2 in saturated $NaClO_4$(Z form). Asterisks indicate the guanine H8 resonances in the syn conformation.

The structural (A, B, and Z forms) dependence of the CIDNP signals of double-stranded nucleic acids was examined. Clear CIDNP signals were observed for A and B form nucleic acids while they were very weak for Z form nucleic acids (Fig. 4). It is suspected that some structural features of Z form (a narrow and deep minor groove , for example) make affinity for flavin low and thus are responsible for these results. Prior to these experiments, the effect of $NaClO_4$ at high concentration on the CIDNP signal of 5′GMP was examined. The intensity of the CIDNP signal observed for the H8 proton of 5′GMP was not affected by $NaClO_4$.

In each case, possible traces of single-stranded nucleic acids as origins of observed CIDNP signals were examined carefully. CD- and UV-temperature profiles indicate that the amount of melted single-stranded materials is completely negligible under the conditions in which the CIDNP experiments were carried out. Nonexistence of single-stranded materials was confirmed further by chemical shift-temperature

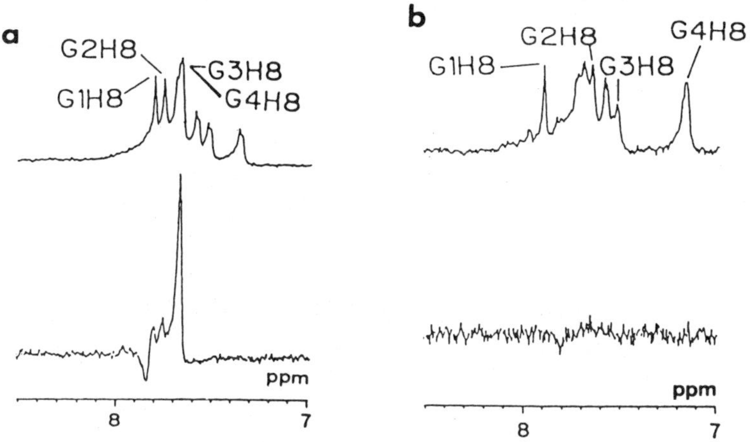

Fig. 5. One-dimensional normal (upper) and photo-CIDNP difference (lower) spectra of (a) d(GGGGCCCC)×2 and (b) the chromomycin-d(GGGGCCCC)×2 complex in 0.15 M NaCl.

profiles of base protons obtained under the same conditions as those for the CIDNP experiments. The CIDNP signals definitely originate from double-stranded nucleic acids.

The drug-free d(GGGGCCCC)×2 shows an intense CIDNP signal (Fig. 5a). When it is bound with chromomycin, however, this signal disappears (Fig. 5b). Prior to this experiment, it was confirmed that the CIDNP signal of 5′GMP is not affected by the addition of chromomycin. Taking into account the structure of the chromomycin-d(GGGGCC-CC)×2 complex discussed in the previous section, the most likely explanation for the disappearance of the CIDNP signal on complex formation would be that: chromomycin bound to the minor groove of d(GGGGCC-CC)×2 interferes with the access of a flavin molecule to the base through direct blocking and/or by inducing structural changes in the DNA, and thus prevents the photo-reaction between a flavin molecule and the base which is essential for the appearance of the CIDNP signal.

The antitumor drug, bleomycin, usually cuts DNA at GpC or GpT sites (*30–32*). d(CCCCAGCTGGGG)×2 (BL-1) is cut at a usual cutting site, GpC, while d(GGGGAGCTCCCC)×2 (BL-2) is cut at an unusual site, GpA, predominantly, although it has a usual cutting site, GpC (Uesugi *et al.*, unpublished results). A CIDNP signal was observed for the H8 proton of the terminal guanosine residue, G12, in the case of BL-1 (Fig. 6a). In the case of BL-2, on the other hand, CIDNP signals were

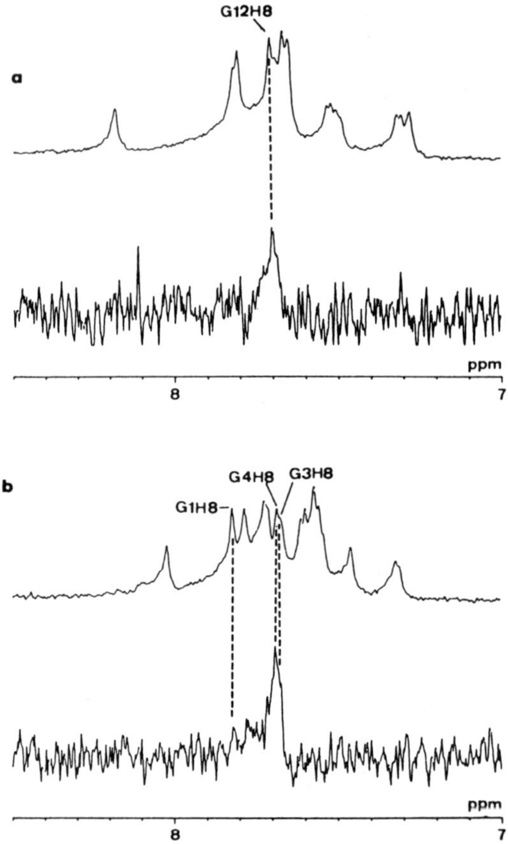

Fig. 6. One-dimensional normal (upper) and photo-CIDNP difference (lower) spectra of (a) d(CCCCAGCTGGGG)×2 (BL-1) and (b) d(GGGGAGCTCCCC)×2 (BL-2) in 0.15 M NaCl.

observed for the H8 protons of G4 and G3 residues, while the CIDNP signal was very weak for the terminal residue, G1 (Fig. 6b). It should be noted that unusual cutting of BL-2 occurs just at the same site where the CIDNP signal was observed selectively. Some structural features responsible for unusual cutting by bleomycin would be reflected on selective observation of the CIDNP signal.

Thus it is suggested that the CIDNP method could be used to detect the interaction between nucleic acids and drugs as well as to study the structures of nucleic acids themselves.

SUMMARY

The structure of putative bent DNA d(GGAAATTTCC)×2 in solution was determined by NMR and restrained molecular dynamics. It shows bending at the junctions between the oligo(dA) tract and the other portions with additional slight bending within the oligo(dA) tract. A new model of DNA bending is proposed on the basis of the NMR observations that the minor groove of the oligo(dA) tract is compressed gradually from 5′ to 3′ along the tract and that the compression almost reaches a plateau level at the third adenine. This model can explain various biochemical data on DNA bending.

The controversial structure of the RNA·DNA hybrid $(rA)_n \cdot d(dT)_n$ was examined by NMR. It is concluded that the rA strand takes on A form while the dT strand takes B form in the duplex.

The structure of the chromomycin-d(GGGGCCCC)×2 complex was determined by NMR. Chromomycin binds as a symmetry-related dimer to the minor groove of the central four residues of the DNA, causing drastic conformational change of the DNA from B form to A form.

The photo-CIDNP signals of double-stranded nucleic acids were observed for the first time. They were observed clearly for A and B form nucleic acids while very weakly for Z form. The CIDNP signal observed for d(GGGGCCCC)×2 disappears when it is bound by chromomycin. The most likely explanation is that chromomycin bound to the minor groove interferes with the access of a flavin molecule to the base and prevents the photo-reaction essential for the appearance of the CIDNP signal. Correlation was found between the unusual cutting of some DNA fragments by the drug, bleomycin, and their CIDNP signals. Thus the possible application of the photo-CIDNP method to structural studies of nucleic acids is indicated.

Acknowledgments
This research was supported by a Grant-in-Aid for Special Project Research from the Ministry of Education, Science and Culture of Japan (#02263103). M.K. was supported by the Human Frontier Science Program for Long-Term Fellowships.

REFERENCES

1. M. Katahira, H. Sugeta, Y. Kyogoku, and S. Fujii: *Biochemistry* **29**, 7214–7222 (1990).
2. M. Katahira, S.J. Lee, Y. Kobayashi, H. Sugeta, Y. Kyogoku, S. Iwai, E. Ohtsuka,

J.M. Benevides, and G.J. Thomas: *J. Am. Chem. Soc.* **112**, 4508–4512 (1990).

3. R. Sakaguchi, M. Katahira, Y. Kyogoku, and S. Fujii: *J. Biochem.* **109**, 317–327 (1991).
4. M. Katahira, H. Sugeta, and Y. Kyogoku: *Nucl. Acids Res.* **18**, 613–618 (1990).
5. M. Katahira, R. Sakaguchi-Katahira, F. Hayashi, S. Uesugi, and Y. Kyogoku: *J. Am. Chem. Soc.* **113**, 8647–8651 (1991).
6. P.J. Hagerman: *Annu. Rev. Biochem.* **59**, 755–781 (1990).
7. D.M. Crothers, T.E. Haran, and J.G. Nadeau: *J. Biol. Chem.* **265**, 7093–7096 (1990).
8. A.V. Fratini, M.L. Kopka, H.R. Drew, and R.E. Dickerson: *J. Biol. Chem.* **257**, 14686–14707 (1982).
9. A. Kintanar, R.E. Klevit, and B.R. Reid: *Nucl. Acids Res.* **15**, 5845–5862 (1987).
10. M. Katahira, H. Sugeta, Y. Kyogoku, S. Fujii, R. Fujisawa, and K. Tomita: *Nucl. Acids Res.* **16**, 8619–8632 (1988).
11. J. Nadeau and D.M. Crothers: *Proc. Natl. Acad. Sci. U.S.A.* **86**, 2622–2626 (1989).
12. A.M. Burkhoff and T.D. Tullius: *Cell* **48**, 935–943 (1987).
13. H.R. Drew and A.A. Travers: *J. Mol. Biol.* **186**, 773–790 (1985).
14. S.C. Satchwell, H.R. Drew, and A.A. Travers: *J. Mol. Biol.* **191**, 659–675 (1986).
15. A. Travers and A. Klug: *Nature* **327**, 280–281 (1987).
16. P.J. Hagerman: *Nature* **321**, 449–450 (1986).
17. H.S. Koo, H.M. Wu, and D.M. Crothers: *Nature* **320**, 501–506 (1986).
18. S.B. Zimmerman and B.H. Pheiffer: *Proc. Natl. Acad. Sci. U.S.A.* **78**, 78–82 (1981).
19. H. Shindo and U. Matsumoto: *J. Biol. Chem.* **259**, 8682–8684 (1984).
20. M. Katahira, Y. Nishimura, M. Tsuboi, T. Sato, Y. Mitsui, and Y. Iitaka: *Biochim. Biophys. Acta* **867**, 256–267 (1986).
21. G. Gupta, M.H. Sarma, and R.H. Sarma: *J. Mol. Biol.* **186**, 463–469 (1985).
22. S.H. Chou, P. Flynn, and B.R. Reid: *Biochemistry* **28**, 2435–2443 (1989).
23. M.W. Van Dyke and P.B. Dervan: *Biochemistry* **22**, 2373–2377 (1983).
24. K. Fox and M.R. Howarth: *Nucl. Acids Res.* **13**, 8695–8714 (1985).
25. X. Gao and D.J. Patel: *Biochemistry* **28**, 751–762 (1989).
26. D.L. Banville, M.A. Keniry, M. Kam, and R.H. Shafer: *Biochemistry* **29**, 6521–6534 (1990).
27. R. Kaptein: In "Biological Magnetic Resonance" (Ed. by L.J. Berliner and J. Reuben), pp. 145–191, Plenum, New York (1982).
28. G.J. Garssen, R. Kaptein, J.G.G. Schoenmakers, and C.W. Hilbers: *Proc. Natl. Acad. Sci. U.S.A.* **75**, 5281–5285 (1978).
29. E.F. McCord, K.M. Morden, A. Pardi, I. Tinoco, and S.G. Boxer: *Biochemistry* **23**, 1926–1934 (1984).
30. A.D. D'Andrea and W.A. Haseltine: *Proc. Natl. Acad. Sci. U.S.A.* **75**, 3608–3612 (1978).
31. M. Takeshita, A.P. Grollman, E. Ohtsubo, and H. Ohtsubo: *Proc. Natl. Acad. Sci. U.S.A.* **75**, 5983–5987 (1978).
32. C.K. Mirabelli, A. Ting, C.H. Huang, S. Mong, and S. Crooke: *Cancer Res.* **42**, 2779–2785 (1982).

Molecular Structure and Life (Ed. by Y. Kyogoku and Y. Nishimura), pp. 67–77, Japan Sci. Soc. Press, Tokyo/CRC Press, Boca Raton (1992)

DNA Bends and Bendability

Donald M. CROTHERS

*Department of Chemistry, Yale University**

I. PERSISTENCE LENGTH AND BENDING ENERGY

The DNA helix axis is straight only under idealized conditions. Bending of the molecule can result from flexing under thermal forces, from intrinsic regularities in the sequence, and from stress induced by binding proteins and other ligands. The major source of DNA bending in cells is protein binding, for example by histone proteins to form polynucleosomes, and the further condensation of these into interphase and metaphase chromosomes. It is also common for regulatory proteins to cause DNA bending at loci such as the promoter sequences preceding transcription start sites.

DNA flexibility can be characterized by the persistence length a, which measures the tendency of a stiff polymer chain to continue in the same direction without bending. Specifically, if θ is the angle between a tangent to the chain axis at its starting end and another tangent a distance L further along the chain contour, then

* New Haven, CT 06511, U.S.A.

$$\langle \cos\ \theta \rangle = e^{-L/a} \tag{1}$$

For a very stiff polymer like double helical DNA, a is large and the average projection $\langle \cos\ \theta \rangle$ of a chain tangent on the starting direction remains substantial until the contour length L is comparable to the persistence length a. It can also be shown by integrating equation (1) that the average projection $\langle z \rangle$ of the chain along its starting direction z is

$$\langle z \rangle = a(1 - e^{-L/a}). \tag{2}$$

Hence the persistence length can be thought of as the average projection $\langle z_\infty \rangle$ of a very long chain ($L \rightarrow \infty$) on its initial direction. The persistence length is about 150 base pairs for DNA of average sequence. It is not presently known how much this parameter varies within the set of naturally occurring DNA sequences.

It requires more energy to bend a stiff polymer chain than a flexible one. Hence one expects the free energy ΔG_{bend} to depend on the persistence length; the relevant equation is

$$\Delta G_{\mathrm{bend}} = \frac{aRT}{2L}(\Delta\theta)^2, \tag{3}$$

where θ is the bend angle as measured by the angle in radians between tangents to the chain before and after the bend. These points are separated by chain contour length L; R is the gas constant and T the temperature. Large bend energies result when substantial curvature occurs over a contour length which is small compared to the persistence length. For example, a 90° bend in 20 bp requires an energy of about 9.3 times RT, or about 5.6 kcal/mol at 300K.

II. INTRINSIC BENDING OF DNA

1. *Sources of Intrinsic Curvature*

Intrinsic bending or curvature of DNA results when certain base sequences or structural motifs are repeated in phase with the DNA helical repeat ($\approx$10.5 bp/turn). In this arrangement the small curvature contributions of the repeating elements add coherently because they are all on the same side of the helix. Among natural sequences, the main contributor to intrinsic DNA curvature is the "A-tract", which consists of homopolymeric dA·dT base pairs, usually about half a helical turn long. An example of a naturally occurring bending sequence is

(5′) AAAAATGTCAAAAAATAGGCAAAAAATGCCAAAAA
which is found in kinetoplast DNA from trypanosome parasites (*1*).

2. Electrophoretic Properties of Bent DNA

Bent DNA molecules are frequently characterized by comparing their electrophoretic mobility with that of normal DNA. It is found empirically that curved molecules migrate more slowly in gels than normal molecules of the same contour length (*1*). This result can be justified by the theoretical equation of Lumpkin and Zimm (*2*), which states that the mobility μ depends on the mean square projection $\langle h_X{}^2\rangle$ of the end-to-end vector of the chain on the field direction, or

$$\mu = \frac{Q}{\xi}\langle h_X{}^2\rangle / L^2\rangle, \tag{4}$$

where Q is the effective charge and ξ the frictional coefficient for translational movement of the chain along a "tube" in the gel. Assuming that the shape of the molecule in the gel reflects its shape in solution, the end-to-end distance, and hence $\langle h_X{}^2\rangle$, should be smaller for a bent molecule than for one that lacks intrinsic curvature. Given that L, Q, and ξ are the same for the two molecules, curvature should cause μ to decrease, as is observed experimentally.

3. Determination of Bend Position by Comparative Electrophoresis

Molecules containing bends differ in shape depending on the position of the bend in the fragment. Placing the bend at the center of the molecule minimizes the mean square end-to-end distance, and therefore minimizes the electrophoretic mobility. Wu and Crothers (*1*) used this principle to identify the A-tract bending sequence shown above. The experimental approach is to compare the mobilities of circularly permuted variants of the same sequence. The mobility data are extrapolated to a maximum at the point where the bending element is at the end of the chain. This approach has been extensively used to characterize the presence and position of DNA bends induced by protein binding (*3*).

4. Determination of Bend Direction by Comparative Electrophoresis

There are two parts to the problem of determining bend directions by comparative electrophoresis (*4*). The first is to establish a direction of bending for the A-tract element, which is then used as a standard in the second part to determine the direction of other bends. The logic by which the direction of A-tract bends was determined began with the observation of Koo *et al.* (*5*) that molecules containing alternating A_6 and T_6 tracts have essentially the same degree of curvature as molecules in which the A_6 tracts are all on one strand and the T_6 tracts on the other. This means that the operation of dyad rotation of an $A_6 \cdot T_6$ tract about an axis running between major and minor grooves at its center, thus interchanging the strands, does not appreciably affect the bend direction. Hence the bend must be directed approximately along the dyad axis. Therefore, its direction is either toward the major or the minor groove in a coordinate frame at or near the center of the A tract.

The remaining ambiguity in A-tract bend direction, whether toward major or minor groove, was resolved by Zinkel and Crothers (*6*), who varied the helical phasing of a segment of repeated A tracts relative to the bend of DNA around catabolite gene activator protein (CAP), which is known to be directed toward the DNA minor groove at the center of pseudo dyad symmetry of the CAP binding site (*7*). Two bends which are "in phase" are in the same direction after an integral number of helical repeats. This leads to a "*cis* isomer", in contrast to the "*trans* isomer" which results when the bends are in opposite directions after an integral number of helical turns. Since the *cis* isomer has the minimum mean square end-to-end distance, it has the smallest mobility of the set of helical phasing isomers, and is easily identified experimentally.

Only the assumption of an A-tract bend toward the minor groove was found to be consistent with the comparative electrophoresis observations. Comparison with additional sequences, such as those studied by Hagerman (*8*) led to a small adjustment of the bend direction, in which the reference coordinate frame is moved about half to one nucleotide away from the center toward the 3′ end of an A tract; the bend direction can be thought of as compressing the minor groove at that position (*4*, *9*). Crystallographic and NMR measurements show that the minor groove is anomalously narrow in A tracts (*4*).

Once the A tract bend direction is established, it can be used to determine the direction of other bends by comparative electrophoresis. For example, Rice and Crothers (*10*) studied multimers made of ligated repeats of the sequences

```
            C G G C G G C G C G G C G T T T T T T G G
“A6”:   C C G C C G C C G C G C C G C A A A A A A
                             A
            C G G C G G C G C G G C G T T T T T T G G
“A11”:  C C G C C G C C G C G C C G C A A A A A A
                   A
```

in which the "bulged" or extra A is located either 6 or 11 nucleotides 3′ to the end of the A tract. The A6 multimers were highly anomalous in electrophoretic mobility, whereas the A11 multimers were normal in mobility. This is the result expected for the indicated combinations of A-tract bends with a bulge bend which is directed away from the strand containing the extra base. The in-phase arrangement in A6 becomes out-of-phase when an additional half helical turn is inserted between the bulge bend and the A-tract bend in A11. Similar experiments were reported for the combination of A-tract bends with the bend produced by the G-G intrastrand crosslink resulting from reaction of *cis*-diamminedichloroplatinum (II) with DNA (*11*).

5. Determination of A-tract Helical Repeat by Comparative Electrophoresis

The electrophoretic anomaly of bent DNA is greatest when the molecules are curved in a plane. This requires an exact match between the sequence repeat of the A-tract elements and the DNA helical repeat. Drak and Crothers (*12*) exploited this logic to determine accurately the helical repeat of DNA molecules in which A_6 tracts alternate with GC rich segments once every helical turn. The maximum anomaly in electrophoretic mobility was found to lie very near a 31 bp sequence repeat, from which it was deduced that the helical repeat is 10.35 bp per turn.

6. Determination of A-tract Bend Angle by Cyclization Kinetics

The theory of gel electrophoresis is not yet sufficiently well developed to enable calculation of absolute curvature from the mobility anomaly. Hence we took recourse to a solution method which has a more rigorous theoretical basis, namely the rate of cyclization of DNA molecules as reported earlier by Shore and Baldwin (*13*, *14*). In this technique one determines the effective concentration of one end of the DNA molecule around the other, called the J value, by comparing the rates of inter- and intra-molecular ligation. We found (*15*) that DNA molecules containing repeated A_6 tracts show maximum cyclization rates at lengths of about 160 bp, compared to about 600 bp for normal DNA molecules (*13*, *14*). The maximum J value for the A-tract containing molecules is about 10^{-4} M, three orders of magnitude larger than the maximum value for normal DNA. Computer simulations of the ring closure process enabled us to conclude that the bend angle is $18° \pm 2°$ per A tract (*15*). This value is in good agreement with our earlier estimate based on rotational dynamics measured by electric dichroism (*16*).

III. PROTEIN-INDUCED DNA BENDS

1. *Sequence Dependence of DNA Bendability*

The bendability of DNA when interacting with proteins depends on base sequence. The rules we have found for this process in the interaction of *Escherichia coli* CAP protein (*17*, *18*) with its binding site are generally similar to those seen for DNA bending in nucleosomal complexes (*19*): A•T rich sequences are preferred when the minor groove is compressed, whereas G•C rich regions are better if the major groove is compressed.

These rules were established experimentally for the CAP-DNA interaction first by comparing the electrophoretic mobility of DNA-CAP complexes in which localized DNA sequence changes were made in the CAP binding site (*17*), excluding the region ~5 base pairs from the dyad axis where the DNA major groove interacts with the helix-turn-helix motif of CAP protein (*7*). This approach takes advantage of the fact that complexes with greater bend angles have lower electrophoretic mobilities. It was found that the sensitivity of bend angle to DNA sequence is greatest at positions ~10 and 16 base pairs from the dyad axis, where first the DNA minor groove and then the major groove turn to face the protein. This finding supports the view that protein-induced DNA bending occurs primarily by roll which narrows the major or minor groove. It was further observed that G•C rich sequences generally gave high mobility (poor bending) and A•T rich sequences gave low mobility (good bending) when the minor groove faces the protein 10–11 base pairs from the dyad axis. At the next roll point, 16 base pairs from the dyad axis, the major groove again faces the protein and the sequence preference is inverted: G•C rich sequences produce stronger bending than A•T rich sequences.

No dependence of bending on base sequence was found at the next potential roll point, 21 base pairs from the dyad axis, from which it seems safe to conclude that CAP-DNA contacts do not extend past 21 base pairs from the dyad axis. However, the observation of sequence-dependent effects 16 base pairs from the dyad axis imply protein-DNA interactions distal to that point, since the energy required to bend DNA must be compensated by favorable protein-DNA interactions on both sides of the bend point.

These effects are also correlated with binding strength, albeit imperfectly: weaker binding generally accompanies weaker bending. This correlation was established by measurement of relative binding affinities of about 50 mutant DNA sequences based on the *lac* promoter CAP binding site (*18*). Correlation of bending and binding strength is not unexpected, since poor bendability at the distal bend points should reduce the tendency of the DNA to bend to take advantage of additional protein contacts to enhance the overall affinity. It is not understood, however, why the correlation is imperfect.

2. Magnitude of the CAP-induced DNA Bend

Crystallographic analysis of the structure of a CAP-DNA complex (*7*) reveals a bend of about 90°. However, while it seems likely that the ~43° kink seen at the major groove on each side of the dyad axis will persist in solution, the path of DNA in the regions distal to these recognition sites is probably influenced by the packing of adjacent molecules in the crystal lattice. Comparative electrophoresis measurements have also been used in an effort to estimate the bend angle (*19*, *20*), but these depend on simplifying assumptions of untested validity. Our approach in this vein (*20*) was to prepare calibration molecules containing both a CAP binding site, placed at the end of the molecule to eliminate the special retardation effects due to bending, and a series of A-tract bends of varying size in the middle of the calibration molecule, to simulate the effects of the bend. The mobility of this set of molecules was compared with the test fragment which had a CAP site at the middle of the molecule, where it was responsible for both the bending effects and the retardation due to simple CAP binding. We concluded that a CAP bend is equivalent to 5.6 A tracts. With each of these contributing 18° of bending, we estimate the CAP bend to be 101°. Determination of this parameter by cyclization kinetic measurements is currently in progress in our laboratory (J.D. Kahn and D.M. Crothers, unpublished).

3. DNA Bending in a Transcriptional Complex at the Lac Promoter

It is of obvious interest to track the fate of the CAP-induced bend through the stages of transcription initiation. In this process the rapidly reversible formation of a "closed" complex in which polymerase encounters the promoter is followed by opening of the DNA helix near the transcription start site:

$$\text{RNA polymerase} + \text{promoter} \overset{K_B}{\rightleftarrows} \text{closed complex} \overset{k_2}{\rightarrow} \text{open complex} \overset{k_i}{\rightarrow} \text{initiation,}$$

$$\text{open complex} \rightleftarrows \text{abortive products}$$

where the vertical arrows indicate the process of abortive cycling in which small RNA chains are initiated and discarded by the polymerase. It has been found that CAP can enhance one or both of the constants K_B and k_2 in exerting its effect to increase the rate of transcription initiation.

Current evidence suggests that the DNA bend induced by CAP protein is maintained and even enhanced in the initiation process up to and including the open complex. DNAse I footprinting assay of the open complex (*22*) showed that the pattern characteristic of an induced bend, in which regions of protection and enhancement of cutting alternate every half helical turn, is at least as pronounced in the open complex containing CAP as in the complex of CAP alone with DNA. It is also found that removal of CAP from the open complex results in an increase in electrophoretic mobility which is too large to be due to loss of the retardation effect of the protein (*23*). Therefore, it is probable that removal of the protein releases a DNA bend, which leads to a strong increase in mobility. Finally, study of the mobility of open complexes containing CAP as a function of the phasing with an additional A-tract bend (*23*) indicated that the bend angle may be increased to 180° or a little more. The result is a working model in which DNA wraps at least halfway around CAP-polymerase in the open complex (*23*).

4. Substitution of A-tract Bends for the CAP Bend

A key question in understanding the function of CAP protein is to deconvolute the effect of the induced DNA bend from the role of possible protein-protein contacts. We and others have addressed this problem by substituting A-tract bends for protein-induced bends. In our experiments, bends resulting from 4 repeated A tracts replaced the natural CAP site in the *lac* promoter (*24*). The phasing of these relative to the promoter was varied in order to assess the significance of bend direction. Using superhelical templates to help activate the *lac* promoter, we found that the rate of open complex formation varied by about 10-fold, depending on the direction of the bend provided by the A tracts. Activity was greatest when the A tract bend was in the same direction as would have been provided by CAP, and nearly undectable when the bend was in the opposite direction. We proposed a model in which the A tracts, with their high degree of curvature, preferentially occupy the loop at the end of an interwound superhelical domain (*24*). This positions the promoter so that DNA naturally wraps around polymerase in the manner suggested by our experiments on the CAP-polymerase-DNA open complex (*23*).

5. Synergistic Protein Binding Mediated by DNA Bending
Two proteins can, in principle, bind cooperatively to DNA if they are able to share the energy cost of inducing a DNA bend which provides favorable DNA contacts for both proteins. Such a synergistic binding mechanism does not require protein-protein contacts, although these are not forbidden either.

The CAP-polymerase complex with DNA may turn out to be an illustration of this principle. Polymerase appears to be activated by DNA conformations which result in looping the DNA upstream of the promoter around the protein. This mechanism requires that there be a polymerase-DNA contact somewhere upstream of the CAP site, so that upstream and downstream polymerase contacts enclose the CAP site in a DNA loop. This configuration pre-bends the DNA so that CAP binding should be enhanced. Similarly, CAP binding bends the DNA to reduce the energy cost of making the upstream polymerase-DNA contacts. The experiments of Straney *et al.* (*22*) suggest a location around -75 to -80 (relative to the start site of transcription) for an upstream polymerase-DNA contact.

In work in progress in our laboratory (J.D. Kahn and D.M. Crothers, to be published), we are exploring the influence on CAP binding due to pre-bending the DNA. The experiments involve determining the relative binding constant of CAP for a linear DNA fragment, compared to binding to the same site incorporated in a small (~160 bp) circle. Because of the small size of the circle, the CAP site is significantly curved even in absence of the protein. Binding enhancement of over two orders of magnitude has been observed due to the pre-bending effect. Hence this mechanism is easily large enough to explain the synergistic interaction between CAP and polymerase in binding promoter DNA. However, this result does not mean that direct CAP-protein contacts are not important in the activation process.

SUMMARY

The DNA helix axis is straight only under idealized conditions. Bending of the molecule can result from flexing under thermal fluctuations, from intrinsic regularities in the sequence, and from stress induced by binding proteins and other ligands. Intrinsic bending or curvature of DNA results when certain base sequences or structural motifs are repeated in phase with the DNA helical repeat (≈10.5 bp/turn). Among natural sequences, the main contributor to intrinsic DNA curvature is the "A tract", which consists of homopolymeric dA•dT base pairs, usually about half a helical turn long. We have concluded that A tracts cause DNA to bend in a direction which is toward the minor groove in a coordinate frame located near the center of an A_{5-6} tract. This result is consistent with simple geometric considerations when the A-tract bend is compared with bends produced by *E. coli* CAP protein, by an extra base on one strand, and by the G-G intrastrand crosslink resulting from reaction of *cis*-diamminedichloroplatinum (II) with DNA. In addition, we have found that each tract bends the helix axis by $18° \pm 2°$.

The bendability of DNA when interacting with proteins depends on base sequence: A•T rich sequences are preferred when the minor groove is compressed, whereas G•C rich regions are better if the major groove is compressed. These effects are also correlated with binding strength, albeit imperfectly. In additional studies with transcription complexes we have found that the bend induced in DNA by CAP protein is maintained and probably even enhanced in the ternary open complex containing CAP and RNA polymerase.

Intrinsic DNA bends can sometimes replace the function of protein-induced bends. The rate of forming open promoter complexes is enhanced as much as ten fold in promoter constructs having the CAP binding site replaced by phased A tracts, provided that the intrinsic DNA bend is in the same direction as the natural bend produced by CAP.

Two proteins can act synergistically in binding to DNA if they both prefer that DNA be bent in the same direction. We have explored the potential magnitude of this effect by measuring the binding affinity of CAP protein for its binding site in small circles, compared to the affinity for linear fragments. Effects of up to two orders of magnitude are observed in some cases.

Acknowledgment

This work was supported by a grant GM 21966 from the National Institutes of Health.

REFERENCES

1. H.-M. Wu and D.M. Crothers: *Nature* **308**, 509 (1984).
2. O.J. Lumpkin and B.H. Zimm: *Biopolymers* **21**, 2315 (1982).
3. D.M. Crothers, M.R. Gartenberg, and T.E. Shrader: *Methods in Enzymol.* **208**, 118

(1991).
4. D.M. Crothers, T.E. Haran, and J.G. Nadeau: *J. Biol. Chem.* **265**, 7093–7096 (1990).
5. H.-S. Koo, H.-M. Wu, and D.M. Crothers: *Nature* **320**, 501–506 (1986).
6. S.S. Zinkel and D.M. Crothers: *Nature* **328**, 178 (1987).
7. S.C. Schultz, G.C. Schields, and T.A. Steitz: *Science* **253**, 1001 (1991).
8. P.J. Hagerman: *Nature* **321**, 449 (1986).
9. H.-S. Koo and D.M. Crothers: *Proc. Natl. Acad. Sci. U.S.A.* **85**, 1763–1767 (1988).
10. J.A. Rice and D.M. Crothers: *Biochemistry* **28**, 4512–4516 (1989).
11. J.A. Rice, D.M. Crothers, A.L. Pinto, and S.J. Lippard: *Proc. Natl. Acad. Sci. U.S.A.* **85**, 4158–4161 (1988).
12. J. Drak and D.M. Crothers: *Proc. Natl. Acad. Sci. U.S.A.* **88**, 3074 (1991).
13. D. Shore, J. Langowski, and R.L. Baldwin: *Proc. Natl. Acad. Sci. U.S.A.* **78**, 4833–4837 (1981).
14. D. Shore and R.L. Baldwin: *J. Mol. Biol.* **170**, 957–981 (1983).
15. H.-S. Koo, J. Drak, J.A. Rice, and D.M. Crothers: *Biochemistry* **29**, 4227–4234 (1990).
16. S.D. Levene, H.-M. Wu, and D.M. Crothers: *Biochemistry* **25**, 3988–3995 (1986).
17. M.R. Gartenberg and D.M. Crothers: *Nature* **33**, 824–829 (1988).
18. D.D. Dalma-Weiszhausz, M.R. Gartenberg, and D.M. Crothers: *Nucl. Acids Res.* **19**, 611–616 (1991).
19. S.C. Satchwell, H.R. Drew, and A.A. Travers: *J. Mol. Biol.* **191**, 659–675 (1986).
20. J.F. Thompson and A. Landy: *Nucl. Acids Res.* **16**, 9687–9705 (1988).
21. S.S. Zinkel and D.M. Crothers: *Biopolymers* **29**, 29–38 (1990).
22. D.C. Straney, S.B. Straney, and D.M. Crothers: *J. Mol. Biol.* **206**, 41–57 (1989).
23. S.S. Zinkel and D.M. Crothers: *J. Mol. Biol.* **219**, 201–215 (1991).
24. M.R. Gartenberg and D.M. Crothers: *J. Mol. Biol.* **219**, 217–230 (1991).

Molecular Structure and Life (Ed. by Y. Kyogoku and Y. Nishimura), pp. 79–91, Japan Sci. Soc. Press, Tokyo/CRC Press, Boca Raton (1992)

Possible Functional Importance of DNA Curvature in *Escherichia coli*

Takeshi MIZUNO

*Laboratory of Microbiology, School of Agriculture, Nagoya University**

The results of recent studies on DNA structure demonstrated the existence of intrinsic (sequence-directed) DNA curvature in free double-stranded DNA molecules (*1*, and refs. therein). The discovery of curved DNA molecules first came through the examination of DNA fragments that displayed unusual electrophoretic behavior on polyacrylamide gels (*2*). DNA molecules exhibiting striking degrees of electrophoretic retardation were originally isolated from restriction digests of kinetoplast minicircle DNA from trypanosomatids (*3*). There are now numerous reports of DNA molecules exhibiting intrinsic curvature in a wide variety of naturally occurring DNA sources (*1*, and refs. therein).

The molecular basis of the origin of such DNA curvatures has been the subject of longstanding debate (*1*). In a number of cases, curved DNA sequences are characterized by the periodical arrangement of short [dA] and [dT] stretches in a given DNA sequence on one face of the DNA helix (*4*). In other words, any short homopolymeric runs of [dA] or [dT] longer than 3 base pairs (bp), when periodically repeated in

* Nagoya, Aichi 464, Japan

approximate phase with the DNA helical repeat, can confer a detectable amount of curvature on an isolated DNA molecule (*4–6*). There are two general classes of models for the origin of intrinsic DNA curvature, namely, the "wedges model" and the "junction model". Trifonov and Sussman proposed that the dinucleotide, [ApA], would open to form wedges through a combination of tilt and roll, and that DNA curvature results from co-orientation of periodically spaced wedges between stacked base pairs (*7*). Koo *et al.* proposed that DNA curvature results from local structural polymorphism in regions of homopolymeric [dA] stretches, and that bending occurs at the junctions between [dA] stretches and adjacent B-form DNA (*4*). In any case, clarification of this particular issue awaits further experiments, including X-ray diffraction studies and NMR studies, as well as theoretical considerations.

Although many examples of naturally occurring curved DNA molecules have been reported, their biological significance is obscure. There have been numerous reports of DNA curvature having potential functions such as replication, recombination and transcription (*1*, and refs. therein). Most of them, however, are laid on simple observations and not on experimental proof, because the problem of establishing the functional importance of each DNA curvature *per se* is rather difficult. Nevertheless, in the last few years, some experimental evidence of the functional significance of DNA curvature has emerged. For example, the requirement of DNA curvature *per se* in both transcription and site-specific recombination was suggested by the results of several experimental efforts (*8*, *9*, and refs. therein). The principal view arising from such studies is that DNA curvature is an important determinant of both the function and the architecture of nucleoprotein complexes concerned in crucial biological processes (*10*, and refs. therein).

To gain general insight into the functional importance of DNA curvature, we have recently taken two different approaches, one is systematic analysis of curved DNA segments randomly cloned from the *Escherichia coli* total genome (*11*), and the other is a search for *E. coli* proteins that preferentially bind to DNA sequences containing curved DNA structures (*12*). The former includes not only cloning of curved DNA segments from the *E. coli* total genome but also clarification of the gene organization in the chromosomal regions surrounding these segments (*13–17*). The latter includes identification of a curved DNA-binding protein (protein H-NS, which has formerly been identified as an *E. coli* histone-like protein), as well as genetic studies on the function of H-NS (*18*, *19*). Here, recent progress emerging from our results is summarized with special reference to the functional importance of DNA curvature.

I. SELECTIVE DETECTION OF CURVED DNA MOLECULES

As described, one hallmark of curved DNA molecules is an anomalously slow electrophoretic mobility on polyacrylamide gel electrophoresis. Findings of curvature in naturally occurring DNA sequences have been primarily based on fortuitous observations of such anomalous electrophoretic mobilities on a one-dimensional polyacrylamide gel. This analytical method by means of one-dimensional polyacrylamide gel electrophoresis, although simple, is inconvenient to convincingly demonstrate the existence of DNA molecules that migrate slower than predicted from their base compositions and sizes (one should somehow first determine their real/actual size). Furthermore, the one-dimensional polyacrylamide gel cannot be applied for a mixture of a large number of DNA segments, simply because of its limited ability of resolution. Although curved DNA molecules are believed to exhibit anomalously slow electrophoretic mobilities, it is also known that they tend to migrate more normally when electrophoresed at higher temperature (>50°C) (*20*). Based on these properties, we devised a two-dimensional polyacrylamide gel electrophoresis method for selective detection of curved DNA segments in a mixture of a large number of DNA segments (*11*). Electrophoresis of a mixture of these was first performed on a cylindrical polyacrylamide gel at a high temperature (60°C). The gel was then subjected to a second slab polyacrylamide gel electrophoresis at a low temperature (<10°C). After the two-dimensional gel electrophoresis, most of the DNA segments applied on the gel were found along a diagonal line. On the other hand, DNA segments containing a curved DNA structure would be expected to migrate off and above the diagonal line, since their electrophoretic mobility is retarded at low temperature.

To evaluate this simple method for the selective detection of curved DNA molecules, a synthetic curved DNA molecule was analyzed in comparison with a synthetic non-curved DNA molecule (Fig. 1). The synthetic molecule, S10–15, consisted of tandem repeats (×15) of a 10 mer [dGGCAAAAACG], which was designed to phase at 10 bp interval $[dA]_5$ stretches in its sequence, whereas the synthetic non-curved sequence, S15–12, consisted of tandem repeats (×12) of a 15 mer [dCCG-GCAAAAACGGGC], designed to phase at 15 bp interval $[dA]_5$ stretches. These two types of DNA segments were analyzed by the

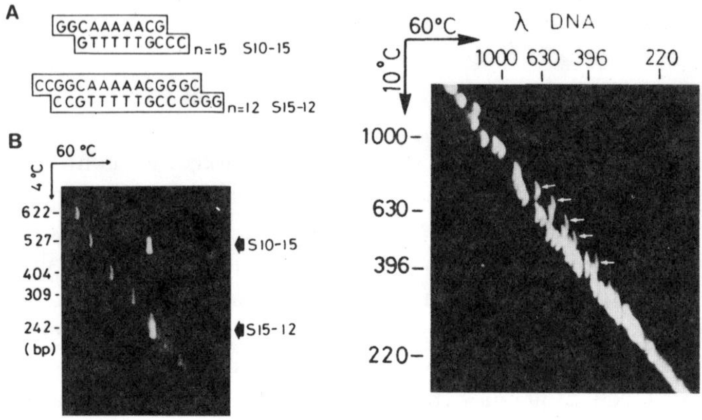

Fig. 1. Selective detection of curved DNA molecules by two-dimensional polyacrylamide gel electrophoresis. Left panel: the 200 bp synthetic curved DNA (S10–15) and the 233 bp synthetic non-curved DNA (S15–12) were analyzed in the presence of the *Hap*II digests of pBR322 by two-dimensional polyacrylamide gel electrophoresis at the two different temperatures indicated. Right panel: bacteriophage lambda DNA was digested by *Hap*II, and then analyzed.

two-dimensional polyacrylamide gel electrophoresis in the presence of the *Hap*II digests of plasmid DNA (pBR322) as an internal control. As shown in Fig. 1, S10–15 migrated above the diagonal line, whereas S15–12 as well as the *Hap*II digests of pBR322 gave spots on the diagonal line. To evaluate this method further, the *Hap*II digests of the bacteriophage lambda total DNA (approximately 50 kb) were subjected to the same analysis (Fig. 1). The majority of DNA segments of lambda phage were found on the diagonal line on the gel, but several spots were clearly seen off and above the diagonal line. These spots exhibiting temperature-dependent anomalous gel electrophoretic mobilities were presumed to be curved DNA segments. It was confirmed that one of them does indeed encompass the lambda replication origin, as previously reported (*21*). It was thus demonstrated that the two-dimensional gel electrophoresis technique can convincingly and effectively separate suspected curved DNA molecules from the majority of normal ones.

II. CLONING OF CURVED DNA SEGMENTS FROM THE *E. COLI* GENOME

In the hope of finding some clues as to the general structural features and the functional roles of curved DNA sequences, we first attempted to

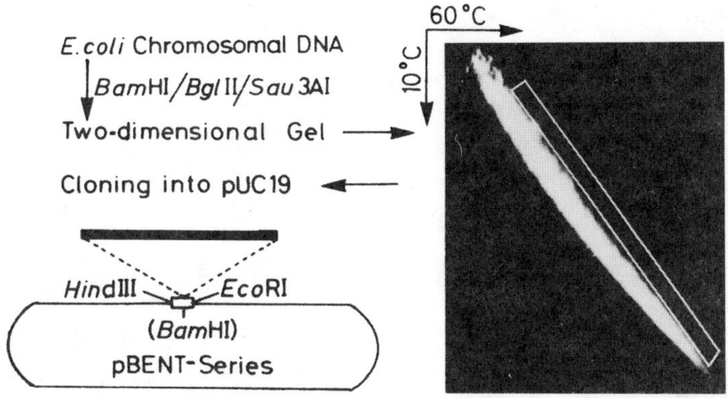

Fig. 2. Strategy for random cloning of curved DNA segments from the *E. coli* total genomic DNA. The white rectangle in the right panel indicates the area of the two-dimensional gel from which DNA was electro-eluted and then cloned onto plasmid pUC19.

randomly clone DNA segments containing a curved DNA structure from the *E. coli* total genome. By applying the method of two-dimensional polyacrylamide gel electrophoresis, this was easily achieved, as schematically summarized in Fig. 2. The *E. coli* total genomic DNA was successively digested with *Bam*HI, *Bgl*II, and *Sau*3AI to yield a mixture of relatively small DNA segments. The restriction fragments were then subjected to two-dimensional gel electrophoresis (Fig. 2). The majority of the digests formed an ethidium bromide stained diagonal line on the gel. However, we assumed that curved DNA segments, which presumably comprise a minor population, must migrate off and above the diagonal line. A piece of gel was cut out from above the diagonal line, and DNAs thus isolated were cloned onto a cloning vector, pUC19. Many of the *Hin*dIII-*Eco*RI segments from the plasmids, which encompass the cloned DNA segments (see Fig. 2), were found to exhibit temperature-dependent anomalous gel electrophoretic mobilities to various extents, as expected. It is known that the anomalous gel mobility of a given curved DNA sequence is dependent upon the position of its bending center relative to the ends of the fragment. This fact immediately suggested that, in order to clone another subset of curved DNA segments, it would be necessary to use another restriction endonuclease to generate a mixture of short DNA segments from the genomic DNA. To make our bank of *E. coli* curved DNAs more comprehensive, we also used *Msp*I, instead of *Sau*3AI. Eventually we obtained a selected set of curved DNA segments (total 27 independent

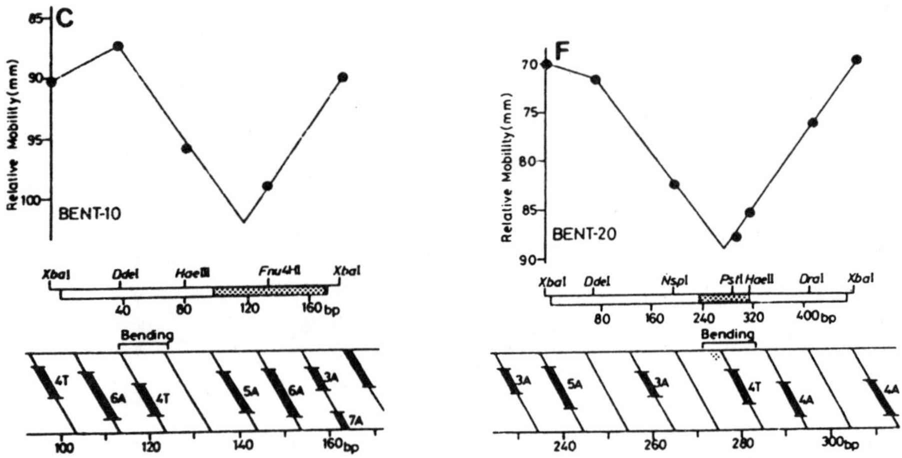

Fig. 3. Circular permutation analysis of the curved DNA segments from *E. coli*. The tandem dimers of BENT-10S and BENT-20S (see Fig. 4D and G, respectively) were digested with the indicated restriction endonucleases, and then analyzed by polyacrylamide gel electrophoresis at 10°C. The relative mobilities of the permutated fragments are plotted against the positions of the restriction sites. The DNA structure of the shaded regions, which contain the determined bending locus were depicted on a cylindrical projection with 10.5 bp per turn, and [dA] and [dT] stretches existing in the regions were plotted. Only one of the strands of the DNA helix is shown for clarity.

clones) from the *E. coli* genome, each of which exhibits an extremely anomalous mobility. We determined their nucleotide sequences. In order to define the sequences responsible for curvature in the respective cloned DNAs, we adopted the circular permutation test developed by Wu and Crothers (*22*). The representative results of this test are presented in Fig. 3. The results of such analyses revealed that in most of the curved DNA segments examined, periodically arranged homopolymeric [dA] and [dT] stretches appeared to be clustered around the nucleotide sequences containing the determined bending centers. These observations on the set of *E. coli* curved DNA sequences are well consistent with the view that short homopolymeric [dA] and [dT] stretches in the DNA helix are the major cause of DNA curvature. The set of curved DNA segments, designated as BENT-series, were subjected to further characterization.

III. CURVED DNA SEQUENCES AND PROMOTER FUNCTION IN *E. COLI*

Although many curved DNA segments were cloned and their nucleotide sequences determined, little information was obtained as to the functional significance of these DNA sequences. Where did these curved DNA segments come from, a coding region of a gene or a non-coding region? What roles do they play in their original positions in the *E. coli* genome? We then determined their exact locations on the *E. coli* genome, and on the basis of the results, carried out further systematic characterization of the curved DNA segments, namely, cloning and sequencing of the genomic regions surrounding them, and a computer-aided search for sequences homologous to them. The origins of 14 curved DNA sequences were thus fairly well identified, as schematically shown in Fig. 4. The results provided us with several pieces of information concerning the gene organization in the genomic regions surrounding the curved DNA segments. As shown, the most striking feature is that many of the curved DNA sequences appear to originate from regions upstream of the coding sequences of the appropriate genes. In addition to these, three of the characterized sequences were found to contain the terminal portion of the insertion sequence IS5 (see Fig. 4 L) (*16*).

Based on the results of statistical analysis of nucleotide sequences for 43 *E. coli* promoters, a positive correlation was previously suggested between promoters displaying a high rate of transcriptional initiation and regions displaying DNA curvature (*23*). In other words, *E. coli* promoters which exhibit high transcriptional initiation rates *in vivo* tend to have high curvature scores for their upstream regions. The experimental evidence presented here clearly supports this view. Furthermore, our results specifically demonstrated that the upstream regions of, for example, the *rrnB*, *proV*, and *argU* genes contain a curved DNA structure. It thus appeared that *E. coli* promoters and their upstream sequences often display an intrinsic DNA curvature.

As previously reported by Bracco *et al.* (*8*), and discussed by Travers (*10*), the curved DNA sequences residing upstream of promoters may act as *cis*-acting transcriptional activators. In this respect, several mechanisms can be speculated to explain the possible effect of DNA curvature on transcription, which is mediated by RNA polymerase. The curved DNA sequences residing upstream of promoters may act as an additional RNA polymerase entry-site, facilitating its binding to the canonical -35 and -10 regions (*24*). The curved regions may bend around RNA polymerase, thus increasing the number of contact sites for RNA polymerase on the DNA (*25*). Recently, McAllister and Achberger demonstrated that the curved DNA sequence, which is located upstream of the promoter of *Bacillus subtilis* phage SP32, allows for better RNA polymerase-promoter interaction (*26*), open complex formation consequently

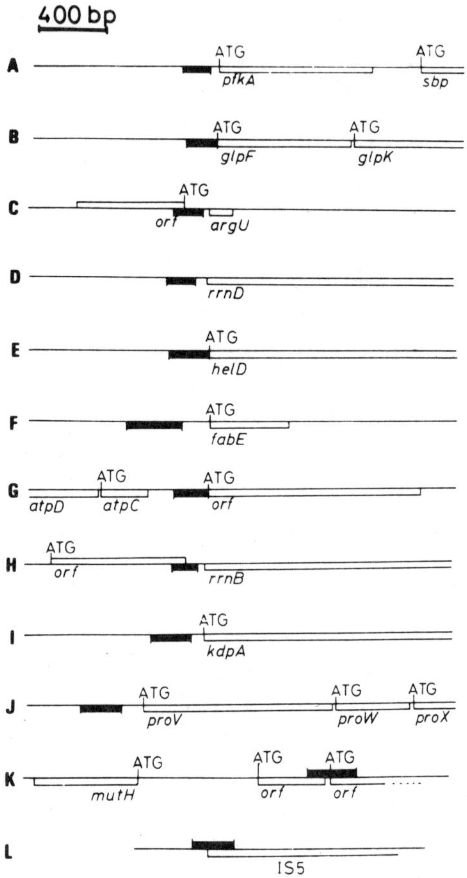

Fig. 4. Schematic representation of the gene organization in the regions of the *E. coli* genome surrounding a set of curved DNA sequences. The positions of the curved DNA sequences are shown by filled rectangles, and the genes located close to the respective curved DNA sequences are indicated by the open rectangles. Note that most of the curved DNA sequences are located at 5′-upstream of the respective genes.

being facilitated. In any event, the results presented here clearly indicate the significance of curved DNA sequences to *E. coli* promoter function. We do not necessarily argue here that curved DNA sequences are exclusively located in non-coding promoter regions in the *E. coli* genome, and at present it is also rather difficult to estimate the frequency of the occurrence of these in a given region of the *E. coli* genome.

IV. CURVED DNA-BINDING PROTEIN FROM *E. COLI*

We then searched for *E. coli* proteins which preferentially bind to a curved DNA sequence even in the presence of an excess amount of non-curved DNA sequence as a competitor. For this purpose, we attempted to purify *E. coli* proteins which preferentially bind to the ideal synthetic curved DNA, S10–15 (see Fig. 1), even in the presence of an excess amount of the competitor synthetic non-curved DNA, S15–12, using gel electrophoretic DNA-binding assay (gel retardation assay). Since these two sequences have nearly the same nucleotide composition, including consecutive $[dA]_5$ stretches, we reasoned that this strategy would allow us to identify proteins which preferentially recognize an overall DNA curvature, sequence-specific DNA-binding proteins (*e.g.*, $[dA]_n$-binding proteins) being excluded. *E. coli* cell-lysate was fractionated by phosphocellulose column chromatography. After incubation of ^{32}P-labeled S10–15 with each fraction in the presence of an excess amount of non-labeled S15–12, the mixtures were subjected to non-denaturing polyacrylamide gel electrophoresis. Several fractions were suspected to contain a protein(s) which may preferentially bind to S10–15. Using successive column chromatography, a protein with an approximate molecular weight of 16,000, which appeared to bind to S10–15 with strong affinity, was purified to near homogeneity. The purified protein, tentatively called 16K, was found to bind to the synthetic curved DNA with strong affinity *in vitro*, although it also binds to the non-curved DNA sequence with lower affinity (Fig. 5). In other words, if the 16K protein can choose between the two DNA sequences, it preferentially recognizes the curved one. Its preferential binding to the curved DNA was further supported by the observation that the non-intercalating DNA-binding ligand distamycin, which is known to remove DNA curvature (*22*), inhibited the binding of the 16K protein to the appropriate sequence (Fig. 5). Since distamycin is believed to interact with the minor groove of adjacent adenine residues and thymine residues on the opposite strand (*27*), [dA] stretches and their narrow minor grooves may be involved, in part, in the preferential binding of the 16K protein to the curved DNA. The major determinant of the observed preferential binding appears to be an overall DNA curvature, rather than sequence-dependent variation in a local DNA conformation caused by a [dA] stretch, since the non-curved DNA used as a negative control should also contain such a topological domain that exhibits an intrinsic micro-curvature.

Since we used a synthetic curved DNA as a probe, the question then

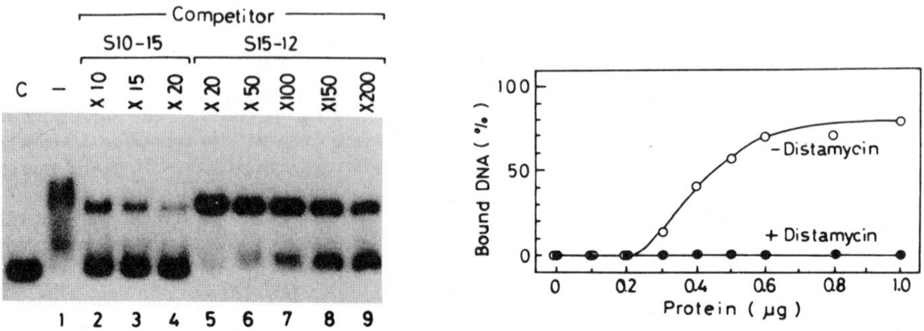

Fig. 5. DNA-binding properties of H-NS. Left panel: purified H-NS was incubated with ^{32}P-labeled S10–15 in the presence of increasing amounts of the indicated non-labeled competitor DNAs. Right panel: ^{32}P-labeled S10–15 was pre-incubated with and without distamycin (5 μM), and further incubated with purified H-NS in the presence of an excess amount of non-labeled S15–12. These samples were subjected to gel retardation assay. In the right panel, the amounts of protein-bound DNA were determined by densitometric quantification of the autoradiograms.

arose as to whether or not the 16K protein binds to naturally occurring curved DNA sequences as well. Gel retardation analyses were carried out for the previously isolated curved DNA sequences encompassing the upstream sequences of the *rrnD*, *helD*, *rrnB*, *kdpA*, and *proV* genes (see Fig. 4). The results clearly showed that the 16K protein exhibits strong affinity for these naturally occurring curved DNA sequences, even when the synthetic or naturally occurring non-curved DNA sequences were present in excessive quantity (data not shown). To provide direct evidence for the preferential binding of the 16K protein to appropriate curved DNA sequences, more precise and extensive DNA-binding experiments should be carried out. The mode of interaction between the 16K protein and its cognate DNA sequence also remains to be elucidated.

V. IDENTIFICATION OF THE CURVED DNA-BINDING PROTEIN

A search for amino acid sequences homologous to that determined for the 16K protein revealed that this protein is most likely identical to the previously identified *E. coli* nucleoid protein, called H-NS (*28*). *E. coli* cells produce a set of small and usually basic proteins that bind to DNA, which are collectively called histone-like proteins because their biochemical properties resemble those of eukaryotic histone (*29*). Of five such *E. coli* proteins so far characterized, H-NS is a neutral protein with a molecular weight of 15,500 (*28*). The gene coding for H-NS has recently been cloned and sequenced (*30*), the deduced amino-terminal amino acid sequence of H-NS was found to be identical to that determined for the 16K protein identified in this study.

The question then arose as to what the physiological function of H-NS is. In this respect, a recent series of suggestive results bearing on the possible function of H-NS was obtained in genetic studies on *drdX*, a regulatory locus involved in thermoregulation of a determinant of uropathogenic pilus adhesion (*31*), and also in genetic studies on *osmZ*, a mutation that causes altered regulation of the *proU* operon in response to an environmental osmotic stimulus (*32*). In these studies, H-NS was demonstrated to be the product of both the *drdX* and *osmZ* loci which are thus allelic. Furthermore, a number of other previously reported mutations, which alter the expression of a variety of unlinked genes, are all allelic (*33*, and refs. therein). Therefore, it is most likely that mutational alterations in H-NS influence the regulation of a subset of unlinked *E. coli* genes and operons at the transcriptional level.

VI. POSSIBLE FUNCTION OF THE CURVED DNA-BINDING PROTEIN, H-NS

The results of extensive genetic studies suggested that H-NS is a global transcriptional regulator. The results of our biochemical studies suggested that it is a DNA-binding protein which preferentially binds, at least *in vitro*, to a variety of curved DNA sequences. These findings together tempted us to propose the view that H-NS may function as a global transcriptional regulator through its ability to preferentially bind to curved DNA sequences, which are often found in regions upstream of *E. coli* promoters. In other words, the curved DNA sequences residing upstream of *E. coli* promoters may function as a *cis*-acting transcriptional regulatory element. The possibility exists that local alterations in the topological DNA curvature may be recognized by H-NS *in vivo*, and that such local alterations in the topological DNA curvature may be affected by changes in physiological conditions, such as growth temperature, intracellular ionic strength and DNA supercoiling. Alternatively, the ability of H-NS to bind to curved DNA sequences may be somehow modulated, *e.g.*, by a modification *in vivo*. In any case, in response to changes under a variety of physiological conditions, H-NS may function either as an activator or a repressor for a variety of unrelated genes through its binding to curved DNA sequences. Verification of this plausible but naive model for the possible function of curved DNA sequences and the cognate recognition protein, H-NS, awaits further experiments.

SUMMARY

There have been numerous reports of DNA curvature having potential functions such as replication, recombination and transcription. Most of them, however, are laid on simple observations rather than on experimental proof. The problem of establishing the functional importance of DNA curvature *per se* is rather difficult. To gain insight into the functional importance of DNA curvature, we have recently taken two different approaches, one is systematic analysis of curved DNA segments randomly cloned from the *E. coli* total genome, and the other is a search for *E. coli* proteins that preferentially bind to curved DNA sequences. The most striking result emerging from the former approach is that many of the curved DNA sequences appear to originate from regions upstream of the coding sequences. On the other hand, the *E. coli* nucleoid protein, H-NS, was demonstrated to preferentially bind to the naturally occurring curved DNA sequences. The results of genetic studies on H-NS demonstrated that mutational alterations in H-NS influence the expression of a variety of unlinked *E. coli* genes. These findings together tempted us to propose the view that the curved DNA sequences residing upstream of *E. coli* promoters may function as a *cis*-acting transcriptional regulatory element, and that H-NS may function as a global transcriptional regulator through its ability to preferentially bind to the curved DNA sequences, which are often found in regions upstream of *E. coli* promoters.

Acknowledgment
This study was supported, in part, by a grant from the Taiko Foundation.

REFERENCES

1. P.J. Hagerman: *Annu. Rev. Biochem.* **59**, 755–781 (1990).
2. J.C. Marini, R. Weisberg, and A. Landy: *Virology* **83**, 254–270 (1977).
3. J.C. Marini, S.D. Levene, D.M. Crothers, and P.T. Englund: *Proc. Natl. Acad. Sci. U.S.A.* **79**, 7664–7668 (1982).
4. H.-S. Koo, H.-M. Wu, and D.M. Crothers: *Nature* **320**, 501–506 (1986).
5. P.J. Hagerman: *Nature* **321**, 449–450 (1986).
6. S. Diekmann: *FEBS Lett.* **195**, 53–56 (1986).
7. E.N. Trifonov and J.L. Sussman: *Proc. Natl. Acad. Sci. U.S.A.* **77**, 3816–3820 (1980).

8. L. Bracco, D. Kotlarz, A. Kolb, S. Diekmann, and H. Buc: *EMBO J.* **8**, 4289–4296 (1989).
9. H.A. Nash: *Trends Biochem. Sci.* **51**, 222–227 (1990).
10. A.A. Travers: *Nature* **341**, 184–185 (1989).
11. T. Mizuno: *Nucl. Acids Res.* **15**, 6827–6841 (1987).
12. H. Yamada, S. Muramatsu, and T. Mizuno: *J. Biochem.* **108**, 420–425 (1990).
13. S. Muramatsu and T. Mizuno: *Nucl. Acids Res.* **17**, 3982 (1989).
14. S. Muramatsu and T. Mizuno: *Nucl. Acids Res.* **17**, 4378 (1989).
15. S. Muramatsu and T. Mizuno: *Mol. Gen. Genet.* **220**, 325–328 (1990).
16. S. Muramatsu, M. Kato, Y. Kohara, and T. Mizuno: *Mol. Gen. Genet.* **214**, 433–438 (1988).
17. K. Tanaka, S. Muramatsu, H. Yamada, and T. Mizuno: *Mol. Gen. Genet.* **226**, 367–376 (1991).
18. H. Yamada, T. Yoshida, K. Tanaka, C. Sasakawa, and T. Mizuno: *Mol. Gen. Genet.*, in press.
19. K. Tanaka, H. Yamada, T. Yoshida, and T. Mizuno: *Agric. Biol. Chem.* **55**, 3139–3141 (1991).
20. S. Diekmann: *Nucl. Acids Res.* **15**, 247–265 (1987).
21. K. Zahn and F.B. Blattner: *Nature* **317**, 451–453 (1985).
22. H.-M. Wu and D.M. Crothers: *Nature* **308**, 509–513 (1984).
23. R.R. Plaskon and R.M. Wartello: *Nucl. Acids Res.* **15**, 785–795 (1987).
24. A.I. Lamond and A. Travers: *Nature* **305**, 248–250 (1983).
25. L. Bossi and D.M. Smith: *Cell* **39**, 643–652 (1984).
26. C.F. McAllister and E.C. Achberger: *J. Biol. Chem.* **264**, 10451–10456 (1989).
27. M.W. van Dyke, R.P. Hertzberg, and P.B. Dervan: *Proc. Natl. Acad. Sci. U.S.A.* **79**, 5470–5474 (1982).
28. K. Spassky, S. Rimsky, H. Garreau, and H. Buc: *Nucl. Acids Res.* **12**, 5321–5340 (1984).
29. D.E. Pettijhon: *J. Biol. Chem.* **263**, 12793–12796 (1988).
30. C.L. Pon, R.A. Calogero, and C.O. Gualerzi: *Mol. Gen. Genet.* **212**, 199–202 (1988).
31. M. Goransson, B. Sonden, P. Nilsson, B. Dagberg, K. Forsman, K. Emanuelsson, and B.E. Uhlin: *Nature* **344**, 682–685 (1990).
32. C.J. Dorman, N. NiBhriain, and C.F. Higgins: *Nature* **344**, 789–792 (1990).
33. C.F. Higgins, J.C.D. Hinton, C.S.J. Hulton, T. Owen-Hughes, G.D. Pavitt, and A. Seirafi: *Mol. Microbiol.* **4**, 2007–2012 (1990).

DNA-BINDING PROTEINS AND THEIR INTERACTIONS

Molecular Structure and Life (Ed. by Y. Kyogoku and Y. Nishimura), pp. 95–101, Japan Sci. Soc. Press, Tokyo/CRC Press, Boca Raton (1992)

Structure and Functional Role of the DNA-binding Protein FIS

Dirk KOSTREWA, Joachim GRANZIN, Daniela C. STOCK, Claus SANDMANN, Jörg LABAHN, and Wolfram SAENGER

*Institut für Kristallographie, Freie Universität Berlin**

Several processes in gene expression require bending of the DNA double helix into topographic arrangements that permit direct interaction and communication of proteins located at distant positions on the DNA. The formation of such synaptic complexes with defined three-dimensional structure is facilitated by certain proteins which bind to specific DNA sequences, thereby bending and folding the DNA into the required form (*1*). One of these DNA binding proteins is the *Escherichia coli* factor for inversion stimulation, FIS. This protein binds to two sites on an enhancer and stimulates the inversion of DNA segments catalyzed by the family of Hin recombinases. This is associated with gene switching in genomes of phages Mu (invertase Gin) and P1 (invertase Cin), and in *Salmonella typhimurium* (invertase Hin) (*2*–4). FIS is also involved in the excisive recombination of phage λ (*5*), it is actively engaged in the transcriptional activation of tRNA and rRNA operons (*6*, *7*), and even in the repression of its own gene (*8*). More recently, FIS was found to be involved in DNA replication by binding to the oriC in *E. coli* (*9*).

* 1000 Berlin 33, Germany

FIS is a small, heat-stable protein of 98 amino acids and occurs as a homodimer. Footprint experiments have shown that it binds to the DNA double helix in two subsequent major grooves (*10*, *11*). The diversity of processes in which FIS is involved suggests that its specific DNA binding site should be degenerate to some extent. It is, in fact, only poorly conserved. The FIS homodimer covers about 25 base-pairs with a loose consensus sequence of the 15 central base-pairs (*12*). When FIS binds to DNA, the latter is bent by about 90° as shown by gel retardation experiments (*13*). All these data suggest that FIS recognizes local DNA structure rather than specific sequence and that its main function is to enhance and to stabilize a local DNA bend in processes where a certain topography is needed (*1*).

FIS not only binds to DNA but also to proteins. This was shown in the Hin-catalyzed inversion which requires formation of a synaptic form called invertasome (*14*). Chemical cross-linking followed by electron microscopy has disclosed that the FIS-bound DNA is in direct contact with the two Hin-bound recombination sites. This means that FIS associates simultaneously with the invertase and with DNA.

I. STRUCTURE OF FIS

The FIS preparation that we used in our experiments was obtained from an overexpression vector (*15*). It was crystallized from phosphate or from polyethylene glycol (*16*). The crystals belong to the orthorhombic space group $P2_12_12_1$ with cell constants $a=47.4$, $b=51.0$, $c=79.7$ Å, and there is one FIS dimer in the asymmetric unit. The structure was determined by a combination of multiple isomorphous replacement methods and solvent flattening. It was refined by constrained least squares at a resolution of 2.0 Å with a final R-factor of 19.2 % for 12,719 X-ray data (*17*, *18*). In an independent study, Dickerson and colleagues arrived at the same structure (*19*).

The amino acid sequence of FIS is shown in Fig. 1. In the X-ray study, we found FIS to be composed of 4 α-helices A, B, C, and D. The first 25 N-terminal amino acids are not seen in the electron density which suggests that they are either dynamically or statically disordered in the crystal lattice. In an attempt to locate the N-terminal tail of FIS at least partially, the mutant Ser8Cys was prepared, and either crystallized and soaked with mercurials or co-crystallized with mercurials. The electron density maps calculated with the X-ray data measured from these crystals did not show any significant peaks that could be associated with

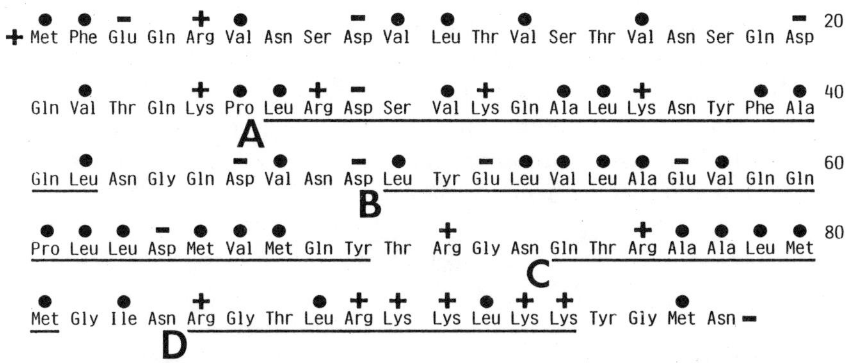

Fig. 1. Amino acid sequence of FIS. The first 25 N-terminal amino acids are not seen in the electron density map. The four α-helices A, B, C, and D are indicated and the respective amino acid sequences are underlined. Positive and negative charges are indicated by (+) and (−), and hydrophobic amino acids by (●). Note that α-helix D carries many positive charges; they probably interact with the DNA major groove and backbone and stabilize the complex.

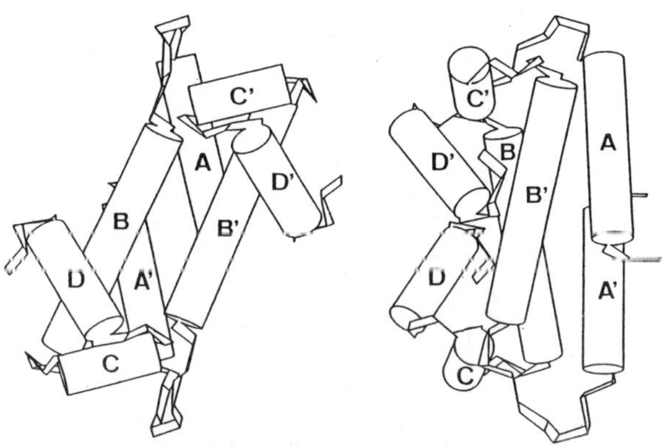

Fig. 2. Front and side views of the FIS dimer. The two molecules of the dimer (indicated by A, B, C, D, and A′, B′, C′, D′) are related by a non-crystallographic dyad symmetry with axis perpendicular to (left) and horizontal within the paper plane (right). α-Helices A, B, A′, B′ form the four-helix-bundle stabilized by hydrophobic forces; C, D, and C′, D′ are helix-turn-helix motifs with α-helices D and D′ in suitable positions to insert into adjacent major grooves of bent B-DNA.

the presence of Hg. This we interpret as being due to dynamic disorder in the crystal lattice.

The structure of the FIS dimer, Fig. 2, shows that helices A, A′, B, B′ of the two monomers are arranged in a four-helix-bundle which is mainly stabilized by hydrophobic forces. Helix B is bent by about 20° at

Cys61. Attached to this four-helix-bundle are helices C, D, and C′, D′ which form a classical helix-turn-helix motif. The motif is superimposable to the helix-turn-helix motifs from repressor molecules (*17–19*). As FIS and repressors have functions that differ widely, this suggests that the helix-turn-helix motif is a module that may be used in a variety of other proteins that interact with DNA.

II. MODEL BUILDING OF A FIS-DNA COMPLEX, MUTATION EXPERIMENTS, AND MUTANT Thr87Ala

Helices D and D′ are exposed to solvent and in the correct position to insert into subsequent major grooves in the DNA double helix. The vertical separation between these two helices, however, is only 25 Å and much shorter than the 29 Å measured between the floors of adjacent major grooves (*17–19*). This suggests that when binding to the FIS dimer, DNA has to bend to bring its two major grooves in contact with helices D and D′. Model building studies have shown that, in fact, DNA can be bent according to these geometric requirements. In our study (*17*, *18*), we have used the DNA with a smooth curvature as known from the

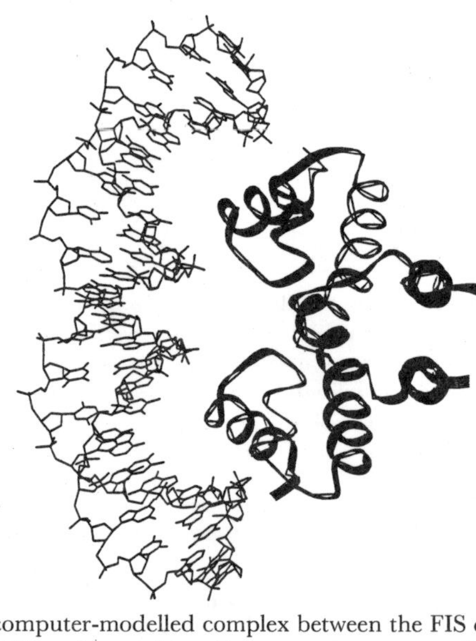

Fig. 3. View of the computer-modelled complex between the FIS dimer and DNA bent by about 90°. Helices D, D′ are inserted in adjacent major grooves, N-termini are to the right of the picture. The structure is twofold-symmetrical with the rotation axis horizontal.

nucleosome octamer and modelled it to FIS as shown in Fig. 3. This is certainly an oversimplification as minor groove sites of DNA become hypersensitive to DNase I digestion when bound to FIS (*11*). This was considered in a model study by Dickerson's group (*19*) who bent DNA by modifying the roll angles in suitable positions to ensure tight contacts between DNA major grooves and FIS helices D and D′.

In a number of mutation experiments (*8*, *20*) it was shown that the function of FIS depends on two local features. One is the helix-turn-helix motif and the other is the N-terminus. If the latter is mutated, this interferes with the proteins involved in the inversion processes and it has only little influence on the DNA binding properties of FIS. On the other hand, if the helix-turn-helix motif or amino acids at the interface between the four-helix-bundle and the helix-turn-helix motif are mutated, this interferes with the binding to DNA and consequently with all the processes in which FIS is involved.

For one mutant of FIS, Thr87Ala, it was shown by gel shift experiments that the bending angle is not 90° but significantly smaller (*8*, *20*). We have crystallized this mutant to see whether the mutation modifies the separation of the helix-turn-helix motifs, thereby changing the bending angle of DNA. It is clear from the X-ray analysis, however, that the structure of the mutant is practically identical with that of wild type FIS. This suggests that it is the interaction between Thr87 and the major groove of DNA that influences the bending angle. Conversely, it appears that the sequence of the FIS binding site on DNA has an influence on the bending angle (*8*, *9*), and that, in fact, few local individual contacts between FIS amino acids and DNA nucleotides are responsible for proper DNA bending.

SUMMARY

The dimeric FIS binds to DNA and bends it by about 90° into a certain topography. It is also involved in DNA excision from phage λ, in the activation of transcription of rRNA and tRNA genes, and in the regulation of the biosynthesis of its own protein. The structure of FIS was determined by X-ray diffraction and refined with data extending to 2.0 Å resolution. Of the 98 amino acids, the first 25 are not seen in the electron density because they are disordered. The rest of the protein is folded into four α-helices. Helices A, A′ and B, B′ form a four-helix-bundle stabilized by hydrophobic interactions, and the attached helices C, D, and C′, D′ form two classical helix-turn-helix motifs. Helices D and

D′ are in suitable positions to interact with DNA, but their separation is about 5 Å shorter than required for binding to two adjacent major grooves in straight DNA. Consequently, DNA has to bend for proper binding. Mutation experiments have shown that the disordered N-terminus is responsible for FIS binding to proteins, and the C-terminus comprising helices C and D is responsible for binding to DNA. The mutant Thr87Ala bends DNA less than wild type FIS but is structurally identical, suggesting that fine details of the FIS-DNA interface are important for proper bending.

Acknowledgments
The authors are grateful to Drs. R. Kahmann and C. Koch for supply of FIS. This work was supported by the Deutsche Forschungsgemeinschaft (Sonderforschungsbereich 344 and Leibniz-Programm), by Bundesministerium für Forschung und Technologie (FKZ 05 413IAB 3), and by Fonds der Chemischen Industrie.

REFERENCES

1. A.A. Travers: *Curr. Opin. Struct. Biol.* **1,** 114–122 (1991); *Curr. Biol.* **1**, 171–173 (1991).
2. R. Kahmann, F. Rudt, C. Koch, and G. Mertens: *Cell* **41**, 771–780 (1985).
3. R.C. Johnson and M.I. Simon: *Cell* **41**, 781–791 (1985).
4. A.C. Glasgow, K.T. Hughes, and M.I. Simon: In "Mobile DNA" (Ed. by D.E. Berg and M.M. Howe), pp. 637–660, American Society for Microbiology, Washington, DC (1989).
5. J.F. Thompson, L.M. de Vargas, C. Koch, R. Kahmann, and A. Landy: *Cell* **50**, 901–908 (1987).
6. L. Nilsson, A. Vanet, E. Vijgenboom, and L. Bosch: *EMBO J.* **9**, 727–734 (1990).
7. W. Ross, J.F. Thompson, J.T. Newlands, and R.L. Gourse: *EMBO J.* **9**, 3733–3742 (1990).
8. R. Osuna, S.E. Finkel, and R.C. Johnson: *EMBO J.* **10**, 1593–1603 (1991).
9. H. Gille, J.B. Egan, A. Roth, and W. Messer: *Nucl. Acids Res.* **19**, 4167–4172 (1991).
10. C. Koch and R. Kahmann: *J. Biol. Chem.* **261**, 15673–15678 (1986).
11. M.F. Bruist, A.C. Glasgow, R.C. Johnson, and M.I. Simon: *Genes Dev.* **1**, 762–772 (1987).
12. P. Hübner and W. Arber: *EMBO J.* **8**, 577–585 (1989).
13. J.F. Thompson and A. Landy: *Nucl. Acids Res.* **16**, 9687–9705 (1988).
14. K.A. Heichmann and R.C. Johnson: *Science* **249**, 511–517 (1990).
15. C. Koch, J. Vandekerckhove, and R. Kahmann: *Proc. Natl. Acad. Sci. U.S.A.* **85**, 4237–4241 (1988).
16. H.-W. Choe, J. Labahn, S. Itoh, C. Koch, R. Kahmann, and W. Saenger: *J. Mol. Biol.* **208**, 209–210 (1989).
17. D. Kostrewa, J. Granzin, C. Koch, H.-W. Choe, S. Raghunathan, W. Wolf, J. Labahn,

R. Kahmann, and W. Saenger: *Nature* **349**, 178–180 (1991).
18. D. Kostrewa, J. Granzin, D.C. Stock, H.-W. Choe, J. Labahn, and W. Saenger: submitted (1992).
19. H.-S. Yuan, S.E. Finkel, J.-A. Feng, M. Kaczor-Grzeskowiak, R.C. Johnson, and R.E. Dickerson: *Proc. Natl. Acad. Sci. U.S.A.* **88**, 9558–9562 (1991).
20. C. Koch, O. Ninnemann, H. Fuss, and R. Kahmann: submitted (1992).

Molecular Structure and Life (Ed. by Y. Kyogoku and Y. Nishimura), pp. 103–113, Japan Sci. Soc. Press, Tokyo/CRC Press, Boca Raton (1992)

Three-dimensional Structure of the DNA-binding Domain of a Protooncogene Product, Myb

Yoshifumi NISHIMURA,*[1] Kazuhiro OGATA,*[1] Hironobu HOJO,*[2] Saburo AIMOTO,*[2] Takahisa NAKAI,*[3] Haruki NAKAMURA,*[3] Akinori SARAI,*[4] and Shunsuke ISHII*[4]

Graduate School of Integrated Science, Yokohama City University,[1] Institute for Protein Research, Osaka University,*[2] Protein Engineering Research Institute,*[3] and Tsukuba Life Science Center, Institute of Physical and Chemical Research (RIKEN)*[4]*

The c-*myb* proto-oncogene is the cellular counterpart of v-*myb* oncogene found in avian myeloblastosis virus (AMV) and E26, the two chicken leukemia retroviruses capable of transforming myeloid cells (*1–2*). The c-*myb* gene product (Myb) binds to DNA in a sequence-specific manner and acts as a transcriptional regulator which is important in maintaining the proliferative state of hematopoietic progenitor cells (*3–8*). Myb protein has three domain structures for DNA-binding, transcription activation and negative regulation from N-terminus to C-terminus as shown in Fig. 1 (*9*). The DNA-binding domain, whose amino acid sequence is highly conserved in various species, consists of three imperfect tandem repeats of 51–52 amino acids (*9–13*). Each repeat contains three conserved tryptophans, spaced 18 or 19 amino acids apart, which play a critical role in sequence-specific DNA-binding (*14–16*). Figure 2 shows the amino acid sequence of the DNA-binding domain of mouse Myb as one example.

*[1] Yokohama 236, Japan
*[2] Suita, Osaka 565, Japan
*[3] Suita, Osaka 565, Japan
*[4] Tsukuba, Ibaraki 305, Japan

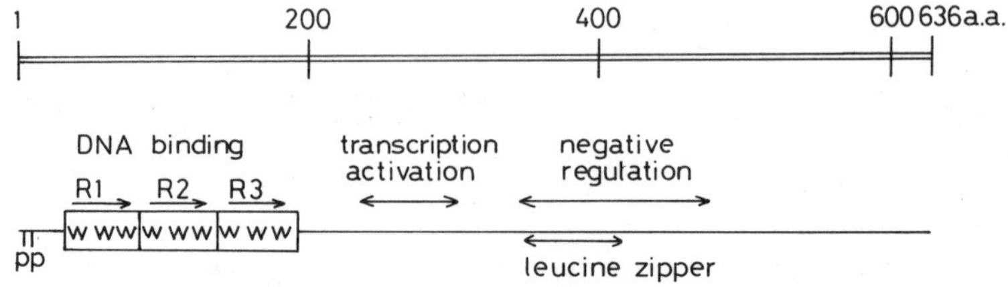

Fig. 1. Functional map of Myb protein. Numbers above the white bar indicate the amino acid sequence numbers of mouse Myb. The domains for DNA-binding, transcription activation, and negative regulation are also indicated. The DNA-binding domain consists of three tandem repeats as shown by the three boxes. Each repeat contains three conserved tryptophans, designated W. The negative regulation domain contains a leucine zipper motif. The N-terminal two serine residues are phosphorylation sites and designated by P. The phosphorylation causes the loss of DNA-binding of Myb.

```
    38 40          50             60             70             80
R1  LGKTRWTREEDEKLKKLVEQNGTDDWKVIANYLPNRTDVQCQHRWQKVLNPE
    90             100            110            120            130            140
R2  LIKGPWTKEEDQRVIKLVQKYGPKRWSVIAKHLKGRIGKQCRERWHNHLNPE
             150            160            170            180            190 193
R3  VKKTSWTEEEDRIIYQAHKRLG-NRWAEIAKLLPGRTDNAIKNHWNSTMRRKV
```

Fig. 2. Comparison of amino acid sequence of each repeat of mouse Myb DNA-binding domain.

The amino acid sequence of each repeat was supposed to be similar to that of homeodomain (*17*), found in various transcriptional regulators responsible for the embryogenesis of *Drosophila*, which contains a helix-turn-helix (HTH) motif (*18*, *19*). Each repeat has therefore been considered to take a similar homeodomain structure (*17*, *20*, *21*). In these three repeats it was thought that the third one recognizes the specific base sequence and the first cooperates in DNA-binding (*15*, *16*). The role of the second repeat is unclear. It was once suggested that it also cooperates in DNA-binding (*15*, *16*), while recently Gabrielsen *et al.* theorized that it binds DNA sequence-specifically based on the results of their site directed mutagenesis experiment (*20*). They assumed that repeats 2 and 3 took a double HTH-related motif. To clarify their roles we investigated the three-dimensional structure of each repeat.

STRUCTURE OF A DNA-BINDING UNIT OF Myb

A polypeptide of the third repeat (R3), corresponding to amino acids 142

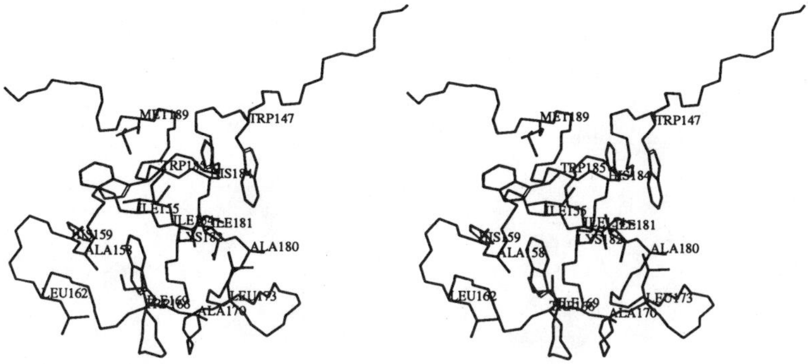

Fig. 3. Stereo view of backbone structure and amino acids forming a hydrophobic core in Myb R3 from distance geometry calculation (*26*).

to 193 of mouse Myb, was chemically synthesized (*22*). Sequence-specific ^{1}H-NMR assignments of R3 were done following the conventional strategy (*23*). At the first stage of distance geometry calculation, geometrically correct structures were generated by the conventional metric matrix algorithm (*24*). Then, following the simulated annealing protocol (*25*), the structures were refined (*26*).

Figure 3 shows the result of the structure calculation. The conformation of the backbone of 147–187 residues was well defined by the NMR data (*26*). It consists of three α-helices (residues 149–162, 166–172, and 178–189) and two turns (residues 163–165 and 173–177). These three helices are maintained by a hydrophobic core formed by three conserved tryptophans, together with two histidines. It is remarkable that Trp's and His's participate in the making of an aromatic hydrophobic core. Helices 2 and 3 appear to form an HTH structure. Root-mean-square-deviations (RMSDs) between the Cα coordinates of 21 residue segments of Myb R3 and of several classical HTH proteins were calculated to be from 1.4 to 1.7 Å (*26*). However, within this group of HTH proteins, RMSD values range within 1 Å (*18*, *27*), while, on the other hand, RMSDs between HTH motif and other protein fragments of the same length taken at random from the protein data bank were mostly larger than 2 Å and rarely around 1.6 Å (*27*). Thus, we conclude that Myb R3 takes a structure closely related to, but distinct from a canonical HTH motif (*26*). This difference is caused by the turn which is one residue longer and contains a proline residue, and also by the difference in the spatial orientation of helices 2 and 3.

The overall molecular architecture of R3 is different from those of

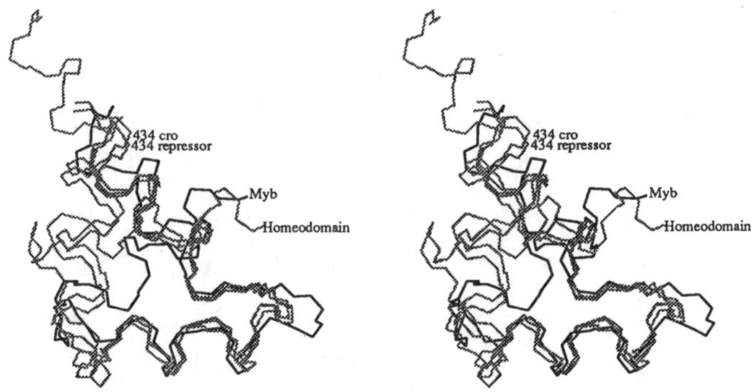

Fig. 4. Stereo view of the DNA-binding domains of Myb R3, 434 cro, 434 repressor and engrailed homeodomain. Only backbone structures are shown.

the engrailed homeodomain (*19*) and DNA-binding domains of 434 repressor (*28*) and 434 cro (*29*) as shown in Fig. 4. In Myb R3, helices 1 and 2 are aligned obliquely with a short turn and helix 3 is almost perpendicular to helix 1, while in the homeodomain helices 1 and 2 are parallel-oriented and in DNA-binding domains of 434 cro and repressor all three helices form a triangular configuration. Thus, Myb R3 takes a new type of DNA-binding structure.

STRUCTURE OF THE DNA-BINDING DOMAIN OF Myb

Based on the present structure of R3 we investigated the structures of the first and second repeats (R1 & R2). As shown in Fig. 2 amino acids forming helices 1 and 2 in R3 are highly conserved in both R1 and R2. This suggests that two corresponding helices are also formed in R1 and R2. Figure 3 shows that three helices of R3 are maintained by a hydrophobic core formed by Trp 147, Ile 154, Ile 155, His 159, Trp 166, Ile 169, Leu 173, Ile 181, His 184, Trp 185, and Met 189 (*26*).

Within these amino acids three tryptophans are conserved, however, two isoleucines (at 154 and 181 positions) and two histidines (at 159 and 184 positions) which act to maintain close contacts with three tryptophans are not conserved, suggesting weaker interactions between the presumed helices in R1 and R2. The detailed comparison of these sequences showed that R1 rather than R2 has a similarity with R3. Especially, the amino acid sequence of the turn region in the HTH-related motif of R3, Leu-Pro-Gly-Arg-Thr-Asp, is not well conserved in R2 (Leu-Lys-Gly-Arg-Ile-Gly) but is highly conserved in R1 (Leu-Pro-

Asn-Arg-Thr-Asp). This suggests that R1 can take an HTH-related structure as R3, but R2 cannot. The overall structure of R1 is believed to be very similar to that of R3, while R2 takes a rather different conformation. Working from this basis we made a model structure of R1 using the following procedure.

(1) First of all, the amino acid sequence of R3 was aligned with that of R1 so as to introduce a gap between Gly 163 and Asn 164 in R3 (see Fig. 2).

(2) Using SYBYL modeling software on graphic workstations , E & S PS330 and IRIS 4D/35GT, the three-dimensional structure of R3 was manipulated; the amino acids of R3 were replaced by those of R1, except for Thr 60 which corresponds to the gap in R3.

(3) The Thr 60 was inserted between Gly 59 and Asn 61, first by excising the Gly 59 and Asn 61, and then by inserting a fragment Gly-Thr-Asn by the loop searching procedure (*30*). The loop conformations which fit the excised gap were sought through the protein data base, and one conformation which causes no van der Waals clashes with surrounding amino acids was selected.

(4) Finally, the structure was energy-refined by the energy minimization procedure in order to avoid steric hindrance.

Figure 5 shows a model of R1 thus obtained. The packing of amino acid sidechains is reasonable to form hydrophobic cores. According to this model, one of the possible differences between structures of R1 and R3 is the stacking between Trp's and His's. In R3, Trp 147 stacks with His 184, and Trp's 166 and 185 interact with His 159. These interactions are lacking in the R1 model; the N-terminal tryptophan, Trp 43, makes

Fig. 5. Stereo view of backbone structure and amino acids which make hydrophobic cores in Myb R1. This model was built by the procedure described in the text.

a

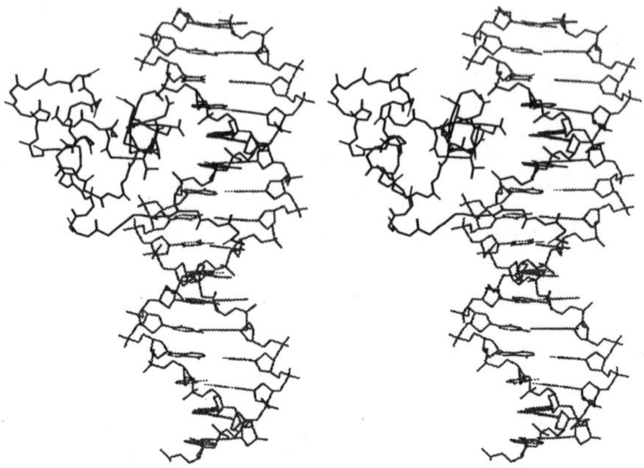

b

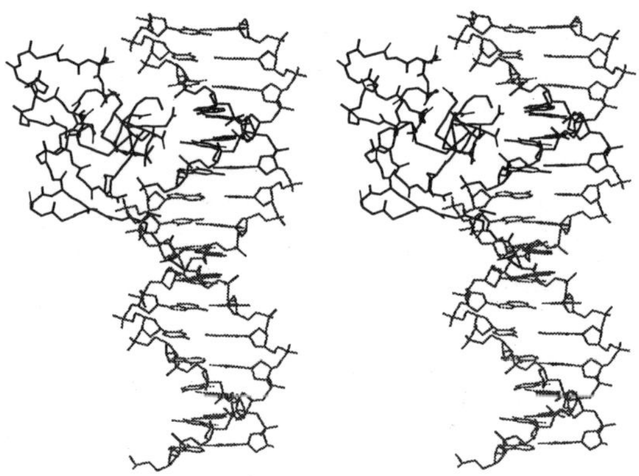

c

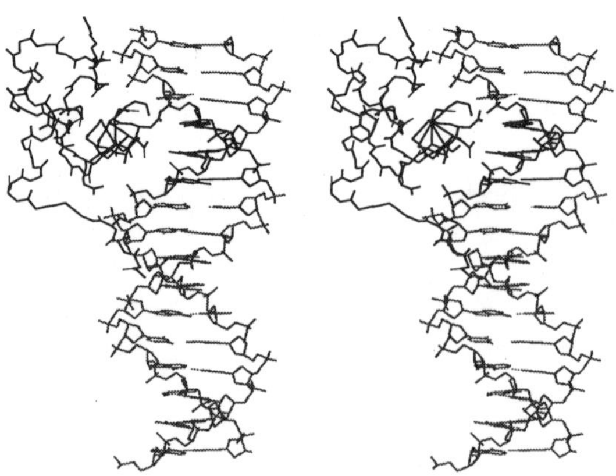

hydrophobic contact with the hydrocarbon-side-chain atoms of Arg 81, while Val 55 in R1 corresponding to His 159 in R3 is a little too far away to make van der Waals contact with Trp 82 and Trp 63. The only histidine, His 80, in R1 seems to be rather exposed in the solvent.

The structures of R3 and R1 raised interesting but bothersome questions: what is the structure of R2 and what is its role for the DNA-binding of Myb ? Perhaps R2 takes a rather different structure, while R1 and R3 retain an HTH-related structure. R2 seems to take a characteristic conformation in order to link R1 and R3, or a part of R2 may serve as a flexible arm found in Matα2 homeodomain (*31*) or λ repressor (*32*). To confirm structures of these repeats we are investigating the DNA-binding domain containing all three repeats by three-dimensional NMR.

INTERACTION BETWEEN DNA AND DNA-BINDING DOMAIN OF Myb

The structural similarity of Myb R3 with homeodomains and bacterial repressors suggests that helix 3 in R3 recognizes the "core" AAC sequence of the consensus Myb binding site, although both repeats 2 and 3 are essential for the sequence specific binding of Myb. We made several models of interaction between Myb R3 and a DNA fragment containing the AAC sequence based on the X-ray structures of engrailed homeodomain-, 434 cro-, and λ repressor-DNA complexes.

When the helices 2 and 3 of Myb R3 are fitted to the corresponding HTH region in the homeodomain-DNA complex (*19*), contacts between amino acids of the helix 3 and base-pairs are poor as shown in Fig. 6a. Part of helix 3 is out of the major groove, and helix 2 cannot make contact with phosphate backbones. Thus, the DNA-binding geometry of the Myb helices 2 and 3 would be quite different from the homeodomain HTH, even though the structure of these two helices of Myb R3 is similar to the homeodomain HTH. On the other hand, Myb R3 fitted to the λ repressor-DNA (*33*) complex forms much better interaction with DNA as shown in Fig. 6b. Helix 2 bridges between phosphate backbones across the major groove, in which Arg 176, Arg 165, Asn 164, Lys 144, and Lys 143 can form electrostatic interactions with phosphate groups. Asn 179 and Lys 182 from helix 3 can form hydrogen bonds with base-pairs, and

←Fig. 6. Stereo views of the model complexes between Myb R3 and DNA. Each model structure was built by superimposing the Myb HTH unit and DNA onto the cocrystal structure of the engrailed homeodomain (a), λ repressor (b) or 434 cro (c).

Thr 177 can form hydrophobic interactions with T's methyl group. However, Asn 183 (which has been suggested to interact with DNA on the basis of results from a mutagenesis experiment), Asn 186 and Ser 187 of helix 3 cannot make direct contact with base-pairs in this geometry. Myb fitted to the 434 cro-DNA complex structure (*34*) is very similar to that fitted to the λ repressor-DNA complex as shown in Fig. 6c.

Some modifications in the relative orientation of Myb R3 from that fitted to λ repressor or 434 cro lead to better interaction with DNA (*26*), where Asn 183, Asn 186, and Ser 187 can make direct contact with base-pairs, and Arg 190 can interact with phosphate group. Asn can form double hydrogen bonds with N6 and N7 of adenosine residue, as observed in the cocrystal structure of DNA fragment and λ repressor (*33*) or the homeodomain (*19*), and is thus a good candidate for forming specific interactions with either A of the AAC sequence. This is also supported by an earlier experiment in which the mutation of Asn 186 into aspartic acid or Lys 182 into glutamic acid resulted in a loss of DNA-binding activity (*16*).

Based on the present model the core "AAC" sequence could be reasonably recognized by R3, although both R2 and R3 are required for the specific binding of Myb. From the R3 structure R2 does not seem to take an HTH structure, so the earlier suggestion that both R2 and R3 take a dual HTH-related structure (*20*) is quite unlikely. If R3 containing the HTH-related structure recognizes AAC sequence from the major groove of DNA as shown in Fig. 6, R2 could not locate at the same position. It is therefore considered that R2 can bind phosphate backbones in the outside region of the AAC sequence to support the recognition of R3.

CHARACTERISTIC OF Myb DNA-BINDING

The DNA-binding domain of Myb has a unique structure containing three tandem repeats. The amino acid sequences of these repeats are relatively conserved, so one might suspect that each repeat might have a similar structure. However, our study suggests that the second and third repeats have different conformations: the third takes an HTH-related structure while the second does not. This feature is different from those of earlier analyzed DNA-binding domains: homeodomain, Cys-His Zn finger, Cys-Cys Zn finger and bZip (*35*). Within bacterial repressors a dimer of DNA-binding domain containing HTH motif binds to DNA with a palindromic sequence. A homeodomain consisting of three helices interacts with DNA by its HTH motif as a monomer with the aid of its N-terminal flexible arm (*31*). A DNA-binding domain containing three Cys-His Zn fingers recognizes the major groove of DNA in the same manner that each finger recognizes a triplet, so that the domain recognizes nine successive base-pairs (*36*). In the glucocorticoid receptor the domain containing two Cys-Cys Zn-nucleated substructures dimerizes in the complex with DNA (*37*).

On the other hand, Myb DNA-binding domain with three repeats binds DNA with AAC sequence. Our model suggests that the third repeat containing an HTH-related structure recognizes the consensus AAC sequence, while for the specific binding of Myb the second one holding a different structure may also be essential. At present we do not know the complex structure between the DNA-binding domain and a specific DNA, however, it is clear that the role of each repeat of Myb DNA-binding domain is different from the others. Myb has a DNA-binding structure unique from other DNA-binding proteins whose structures have been established.

SUMMARY

The DNA-binding domain of a protooncogene product, Myb consists of three imperfect tandem repeats with 51–52 amino acids. We investigated the structure of the third repeat by NMR and found that it takes a structure which includes three α-helices maintained by a hydrophobic core formed by three tryptophans, together with two histidines. Helices 2 and 3 form a structure related to but distinct from a canonical helix-turn-helix motif. In particular, the turn between these helices is one amino acid longer than the corresponding turn in bacterial repressors and homeodomains, and contains a proline residue. Based on the identified structure of the third repeat, it is considered that the first repeat has a similar but less stable structure containing a helix-turn-helix-related motif, while the second repeat has a different conformation. The apparent similarity of amino acid sequences of the three repeats may not be correlated to the three-dimensional structure of each repeat. The second and third repeats are essential for the sequence-specific binding of Myb, but the role of each repeat is believed to be different from the others.

Acknowledgments

We thank Dr. C.O. Pabo for giving us three-dimensional coordinates of the engrailed homeodomain X-ray structure. This work was supported by

Yokohama City University and special-grants-in-aid (Nos. 02263104, 03259104) from the Ministry of Education, Science and Culture of Japan.

REFERENCES

1. M. Roussel, S. Saule, C. Lagrou, H. Beug, T. Graf, and D. Stehlin: *Nature* **281**, 452–455 (1979).
2. K.-H. Klempnauer, T.J. Gonda, and J.M. Bishop: *Cell* **31**, 453–463 (1982).
3. H. Biedenkapp, U. Borgmeyer, A.E. Sippel, and K.-H. Klempnauer: *Nature* **335**, 835–837 (1988).
4. H. Nakagoshi, T. Nagase, C. Kanei-Ishii, Y. Ueno, and S. Ishii: *J. Biol. Chem.* **265**, 3479–3483 (1990).
5. Y. Nishina, H. Nakagoshi, F. Imamoto, T.J. Gonda, and S. Ishii: *Nucl. Acids Res.* **17**, 107–117 (1989).
6. K. Weston and J.M. Bishop: *Cell* **58**, 85–93 (1989).
7. S.A. Ness, A. Marknell, and T. Graf: *Cell* **59**, 1115–1125 (1989).
8. M.L. Mucenski, K. McLain, A.B. Kier, S.H. Swerdlow, C.M. Schereiner, T.A. Miller, D.W. Pietryga, W.J. Scott, and S.S. Potter: *Cell* **65**, 677–689 (1991).
9. H. Sakura, C. Kanei-Ishii, T. Nagase, H. Nakagoshi, T.J. Gonda, and S. Ishii: *Proc. Natl. Acad. Sci. U.S.A.* **86**, 5758–5762 (1989).
10. K.-H. Klempnauer and A.E. Sippel: *EMBO J.* **6**, 2719–2725 (1987).
11. C.W.B. Peters, A.E. Sippel, M. Vingron, and K.-H. Klempnauer: *EMBO J.* **6**, 3085–3090 (1987).
12. J. Paz-Ares, D. Ghosal, V. Wienand, P.A. Peterson, and H. Saedler: *EMBO J.* **6**, 3553–3558 (1987).
13. K. Tice-Baldwin, G.R. Fink, and K.T. Arndt: *Science* **246**, 931–935 (1989).
14. I.A. Anton and J. Frampton: *Nature* **336**, 719 (1988).
15. C. Kanei-Ishii, A. Sarai, T. Sawazaki, H. Nakagoshi, D.-N. He, K. Ogata, Y. Nishimura, and S. Ishii: *J. Biol. Chem.* **265**, 19990–19995 (1990).
16. P. Saikumar, R. Murall, and E.P. Reddy: *Proc. Natl. Acad. Sci. U.S.A.* **87**, 8452–8456 (1990).
17. J. Frampton, A. Leutz, T.J. Gibson, and T. Graf: *Nature* **342**, 134 (1989).
18. Y.Q. Qian, M. Billeter, G. Otting, M. Muller, W.J. Gehring, and K. Wüthrich: *Cell* **59**, 573–580 (1989).
19. C.R. Kissinger, B. Liu, E. M.-Blanco, T.B. Kornberg, and C.O. Pabo: *Cell* **63**, 579–590 (1990).
20. O.S. Gabrielsen, A. Sentenac, and J. Frampton: *Science* **253**, 1140–1143 (1991).
21. J. Frampton, T.J, Gibson, S.A. Ness, G. Doderlein, and T. Graf: *Protein Engineering*, **4**, 891–911 (1991).
22. H. Hojo and S. Aimoto: *Bull. Chem. Soc. Jpn.* **64**, 111–117 (1991).
23. K. Wüthrich: "NMR of Proteins and Nucleic Acids," Wiley, New York (1986).
24. G.M. Crippen: In "Distance Geometry and Conformational Calculations" (Ed. by D. Bawden) Chemometrics Research Studies, Vol. 1, Wiley, New York (1981).
25. P.C. Driscoll, A.M. Gronenborn, L. Beress, and G.M. Clore: *Biochemistry* **28**, 2188–2198 (1989).

26. K. Ogata, H. Hojo, S. Aimoto, T. Nakai, H. Nakamura, A. Sarai, S. Ishii, and Y. Nishimura: *Proc. Natl. Acad. Sci. U.S.A.*, **89**, 6428–6432 (1992).
27. R.G. Brennan and B.W. Mathews: *J. Biol. Chem.* **264**, 1903–1906 (1989).
28. A. Mondragon, S. Subbiah, S.C. Alamo, M. Drottar, and S.C. Harrison: *J. Mol. Biol.* **205**, 189–200 (1989).
29. A. Mondragon, C. Wolberger, and S.C. Harrison: *J. Mol. Biol.* **205**, 179–188 (1989).
30. T. Blundel, D. Carney, S. Gardner, F. Hayes, B. Howli, T. Hubbard, J. Overington, D.A. Singh, B.L. Sibanda, and M. Sutcliffe: *Eur. J. Biochem.* **172**, 513–520 (1988).
31. C. Wolberger, A.K. Vershon, B. Liu, A.D. Johnson, and C.O. Pabo: *Cell* **67**, 517–528 (1991).
32. N. Clarke, L.J. Bemer, H.R. Goldberg, C. Berkower, and C. Pabo: *Science* **254**, 267–270 (1991).
33. S.R. Jordan and C.O. Pabo: *Science* **242**, 893–899 (1988).
34. C. Wolberger, Y. Dong, M. Ptashne, and S.C. Harrison: *Nature* **335**, 789–795 (1988).
35. S.C. Harrison: *Nature* **353**, 715–719 (1991).
36. N.P. Pavletich and C.O. Pabo: *Science* **252**, 809–817 (1991).
37. B.F. Luisi, W.X. Xu, Z. Otwinowski, L.P. Freedman, K.R. Yamamoto, and P.B. Sigler: *Nature* **352** 497–505 (1991).

Molecular Structure and Life (Ed. by Y. Kyogoku and Y. Nishimura), pp. 115–127, Japan Sci. Soc. Press, Tokyo/CRC Press, Boca Raton (1992)

Molecular Recognition in the Homeodomain-DNA System

Kurt WÜTHRICH,*[1] Gottfried OTTING,*[1] Yan Qiu QIAN,*[1] Martin BILLETER,*[1] and Walter GEHRING*[2]

Institut für Molekularbiologie und Biophysik, Eidgenössische Technische Hochschule-Hönggerberg[1] *and Biozentrum der Universität Basel, Abt. Zellbiologie*[2]

The genetic control of development is based upon the coordinated and concerted action of thousands of genes in a precise spatial and temporal pattern, which requires regulatory circuits in which developmental master control genes regulate the activity of large groups of subordinate genes. *Drosophila* geneticists identified a class of so-called *homeotic mutations* in genes, which might represent such master control genes. Homeotic mutations in *Drosophila* are generally defined as mutations which lead to the transformation of structures of one body segment into the corresponding (homologous) structures of another segment. For example, the homeotic mutant called *Antennapedia (Antp)* transforms the antennae on the head of the fly into a pair of legs.

It was not until the advent of gene cloning that the structure and function of homeotic genes became amenable to analysis at the molecular level, and that their role as master control genes specifying the body plan could be unequivocally demonstrated. Both the *bithorax*-complex (*1*) and the *Antp*-complex (*2*, *3*) were cloned by chromosomal walking, and the transcription units corresponding to individual homeotic genes were

*[1] CH-8093 Zürich, Switzerland.
*[2] CH-4056 Basel, Switzerland.

identified. In the course of these studies it was discovered that homeotic genes share a characteristic short DNA segment, the *homeobox* (*4–6*), that encodes a precisely defined domain of the corresponding protein, the *homeodomain.*

The significance of the homeobox homology was demonstrated in *Drosophila* by using the *Antp* homeobox as a probe to isolate additional homeotic genes which had previously been identified by genetic techniques (*4*), as well as other developmental control genes, such as the segment polarity gene *engrailed* (*en*) (*7*). Furthermore, an evolutionary survey showed that the homeobox is found not only in *Drosophila* and other arthropods, but quite unexpectedly also in vertebrates (*8–10*). Although vertebrates differ in fundamental ways from *Drosophila* in their mode of development and their body plan, there is a remarkable degree of sequence conservation in the homeodomains. For example, the human *c1* (=*Hox 2.C*) gene (*11*) encodes a homeodomain which is identical with the *Antp* homeodomain of *Drosophila* in 59 out of 60 amino acid residues. This raises the possibility that vertebrates and invertebrates may use the same classes of genes to specify fundamental aspects of their body plans.

A diagram indicating the frequency with which particular amino acids are found at given positions in the homeodomains of different species is presented in Fig. 1. There are the 4 invariant amino acid residues Trp 48, Phe 49, Asn 51, and Arg 53 and the 14 additional highly conserved residues Arg 5, Gln 12, Leu 16, Glu 17, Phe 20, Tyr 25, Arg 31, Ala(or Ser) 35, Leu 38, Leu 40, Ile(or Val) 45, Arg 52, Lys(or Arg) 55, Lys(or Arg) 57. The structural significance of the conserved residues will be discussed below. The consensus sequence as determined from Fig. 1 is most closely related to the *Antennapedia* sequence. Therefore, *Antp* may reflect the prototype or ancestral sequence.

A computer search through protein data banks revealed some distant sequence homologies between homeodomains and the yeast mating type proteins (*12*). Although the sequence identity between *Mat a1* and $\alpha 2$ of yeast and the *Antp* homeodomain of *Drosophila* is rather low (about 25%), the homology appears significant. In particular, three and four of the amino acids that are invariant in all homeodomains described so far are also present in yeast *Mat a1* and $\alpha 2$. For the *Mat a1* protein it had been proposed from sequence comparisons and some other structural considerations that it might contain a helix-turn-helix motif similar to the one found in the prokaryotic gene regulatory proteins *lac*, *cro*, *CAP*, and λ-repressor, on the basis of NMR or X-ray crystallographic studies (*13*). Since the sequences encoding the putative helix-turn-helix motif in

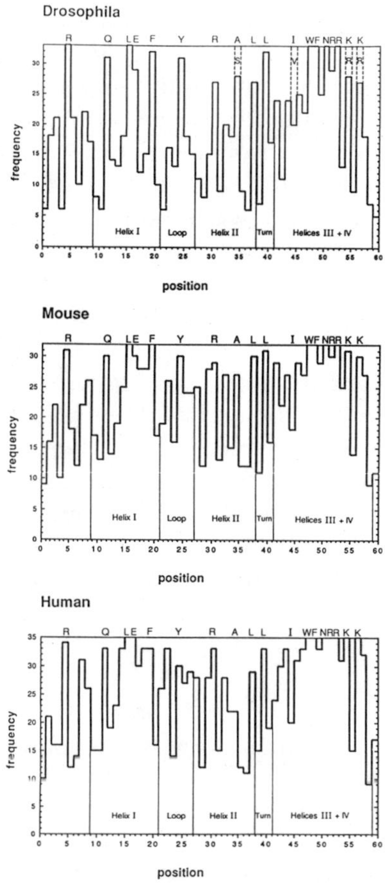

Fig. 1. Comparison of the amino acid sequences of homeodomains from *Drosophila*, mouse and man. For each of the sequence positions 1–60 the height of the bar indicates the number of times the most frequently encountered amino acid is found among the 33, 32, and 35 homeodomain sequences determined to date for *Drosophila*, mouse and man, respectively. At positions 48, 49, 51, and 53 all known homeoboxes encode the same amino acid. These invariant residues and the 14 most highly conserved amino acids are labelled using the one-letter code. As indicated for the *Drosophila* sequences, positions 35, 45, 55, and 57 are exclusively occupied by one of a pair of amino acid types, *i.e.*, position 35 is occupied either by Ala or Ser, 45 by Ile or Val, 55 by Lys or Arg, and 57 by Lys or Arg. The locations of secondary structure elements (α-helix, loop, turn) are also indicated.

Mat a1 are those which are homologous to the homeodomain, there was then the indication that homeodomains might also contain a helix-turn-helix motif (*12*, *14*), although the sequence homology between the *Antp* homeodomain and the corresponding amino acids in the procaryotic helix-turn-helix proteins is not statistically significant (*15*).

The sequence determinations and comparisons have by now been followed up by experimental structure determinations. For the free proteins complete three-dimensional structures in solution were obtained for the *Antp* homeodomain (*16*, *17*) and the *Antp*(*C39→S*) homeodomain (*18*), and a secondary structure was reported for *Mat* α2 (*19*). The NMR solution structure of a complex formed between the *Antp*(*C39→S*) homeodomain and a 14-base pair DNA duplex including the BS2 binding site was initially solved using ^{15}N-labeling of the protein (*20*) and has since been refined using ^{13}C-labeling of the homeodomain (unpublished), and X-ray crystal structures were reported for DNA complexes of the *engrailed* homeodomain (*21*) and *Mat* α2 (*22*). In this article we shall use the NMR structures determined in our laboratory (*16*–*18*, *20*) as a basis for discussions on the nature of the protein-DNA interactions. To the level of structural detail used here, the subsequently solved crystal structures of homeodomain-DNA complexes (*21*, *22*) are in full agreement with the conclusions from the NMR structures.

I. THE SOLUTION STRUCTURE OF THE *Antp* HOMEODOMAIN

A 68-residue recombinant polypeptide was used for the structure determination of the *Antp* homeodomain in aqueous solution. To prevent oxidative dimerization of the protein (*23*), the cysteinyl residue at position 39 was replaced by serine. This mutation does not affect the three-dimensional structure of the protein (*16*–*18*). Figure 2 shows a superposition of the polypeptide backbone in the twenty NMR conformers of the *Antp* (*C39→S*) homeodomain used to represent the solution structure. All these conformers are compatible with the NMR data with only minor constraint violations, and the narrow width of the bundle for the polypeptide segment from residues 7–59 is characteristic of a high quality NMR structure (*24*). This well defined core of the *Antp*(*C39→S*) homeodomain structure, which is indistinguishable from that of the native *Antp* homeodomain (*16*, *17*), consists of helix I from residues 10–21, a connecting loop of residues 22–27 leading to helix II (residues 28–38), a tight turn of residues 39–41, and a third helical region from residues 42–59. This latter region can be divided into a well defined α-helix III including the residues 42–52, and a more disordered and flexible helix IV containing residues 53–59. The flexible chain ends of residues 0–6 and 60–67 (not shown) stick out into the solvent at the top and on the extreme left in the

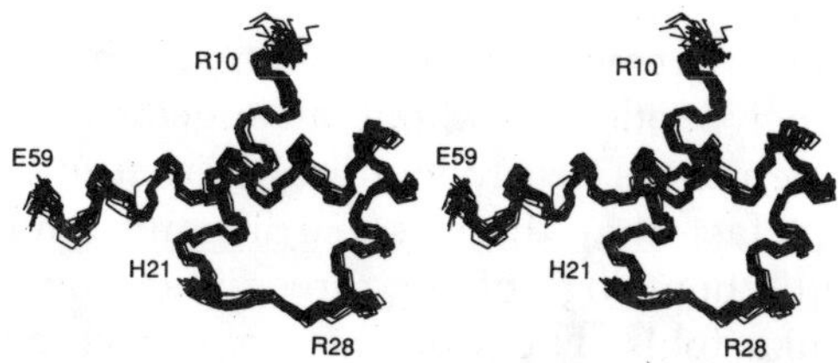

Fig. 2. Superposition for optimal fit (*i.e.*, minimal pairwise root mean square deviation) of 20 *Antp(C39→S)* homeodomain conformers used to represent the solution structure of the protein. Each conformer was obtained from the same NMR input data using distance geometry calculations and subsequent restrained energy refinement. A stereo view of the bonds connecting the backbone atoms N, C^{α}, and C' of the residues 7–59 is shown (reproduced from ref. *18*).

orientation of the molecule in Fig. 2. Helices I and II are aligned in almost exactly antiparallel fashion, and the helices III and IV are arranged approximately perpendicular to the other two helices. Helix II, helix III, and the connecting turn form a typical helix-turn-helix motif (*16*).

In the homeodomain molecular architecture, helices I, II, and III are held together by a core of 11 hydrophobic amino acids with tightly packed side chains. Six of these core amino acids are either invariant (Trp 48, Phe 49) or very highly conserved (residues 16, 38, 40, and 45) in a wide variety of homeodomains (ref. *25*; Fig. 1). Since the hydrophobic core of a protein plays a key role in determining its structure and since most of the conserved amino acids form the core, one can anticipate that different homeodomains fold into similar three-dimensional structures. This prediction has been confirmed by all structure determinations reported so far, *i.e.*, the secondary structure of free *Mat* α2 in solution (*19*) and the DNA-bound *engrailed* homeodomain (*21*) and *Mat* α2 (*22*) in crystals. All additional highly conserved side chains, in particular the invariant residues, Asn 51 and Arg 53 and the highly conserved residues Lys 55 and Lys 57, are hydrophilic and most of them are part of the helices III and IV, where they may be involved in intermolecular interactions in the complexes with DNA.

II. COMPARISON OF HOMEODOMAINS WITH PROCARYOTIC DNA-BINDING PROTEINS

DNA-binding domains of procaryotic DNA-binding proteins that contain a helix-turn-helix motif and for which a three-dimensional structure was determined either by crystallography or by NMR include the *434* repressor (*26*), λ-repressor (*27*), *434 cro* protein (*28*), *trp* repressor (*29*),

CAP (*30*), and *lac* repressor (*31–33*). Comparisons with these proteins showed that the polypeptide backbone arrangement in the region of the helix-turn-helix segment is nearly identical to that in the *Antp* homeodomain, despite the fact that there is hardly any amino acid sequence homology and little homology of the three-dimensional structures in the rest of the molecular folds. The propensity of conserved lysyl and arginyl residues along the helices III and IV in the *Antp* homeodomain would support that these helices function as the DNA recognition sites, in analogy to the procaryotic systems. However, considering that there is very little homology between the amino acid sequences or between the three-dimensional structures outside the helix-turn-helix segment, these structure comparisons could not conclusively answer the question to what extent the DNA binding mode might be similar or different for the procaryotic proteins and the homeodomains.

III. HOMEODOMAIN-DNA COMPLEXES

1. *Stability of Homeodomain-DNA Complexes*

The DNA-binding properties of homeodomain-containing proteins were first studied *in vitro* with fusion proteins to bacterial β-galactosidase (*34*, *35*), and then with full length proteins produced in bacteria (reviewed in ref. *36*). The DNA binding sites turned out to be very A=T rich (*25*, *37*). DNA-binding properties of the *Antp* homeodomain were studied *in vitro* using a homogeneous preparation of the recombinant homeodomain and synthetic oligodeoxynucleotide duplexes (*23*, *38*). No functional *in vivo* binding sites are known for the *Antp* protein. However, the high affinity binding sites BS1, BS2, and BS3 for the homeodomain-containing *ftz* protein, which are located near *ftz*-regulated genes, are also high affinity binding sites for the *Antp* protein (*23*). Since *ftz* contains a homeodomain with a closely similar three-dimensional structure as the *Antp* homeodomain (unpublished data), we used these binding sites as target sequences for our *in vitro* studies. The *Antp* homeodomain binds with a 1 : 1 stoichiometry to these binding sites, with equilibrium dissociation constants of about 10^{-9} M (*38*). This is in contrast to procaryotic helix-turn-helix containing proteins, which bind as dimers to palindromic operator sites (*39*) and have greatly reduced binding affinities for the monomeric DNA-binding domains with their half-operator sites (*38*). The molecular basis for the high stability of the homeodomain-DNA complexes is provided by additional contacts with the DNA, in particular in the minor groove (see following section). However, it should also be emphasized that these *in vitro* studies do not rule out the possibility that *in vivo* the complete *Antp* protein might bind as a multimer, or form heterooligomers with other DNA-binding proteins or transcription factors.

2. The NMR Structure of an Antp(C39→S) Homeodomain Complex with a 14-Base Pair DNA Duplex

Sequence-specific NMR assignments were established for both the protein and the DNA in a complex of the *Antp*(*C39*→*S*) homeodomain and a 14-base pair DNA duplex containing the *BS2* binding site (*20*). The structural information from the NMR data obtained with the ^{15}N-labeled homeodomain is summarized in Fig. 3. The first important result, ascertained by several independent lines of experimental evidence, showed that the conformations of the homeodomain and the DNA are very similar for the molecules in the complex and free in solution. Thus, with few exceptions the same short *intramolecular* proton-proton distances are

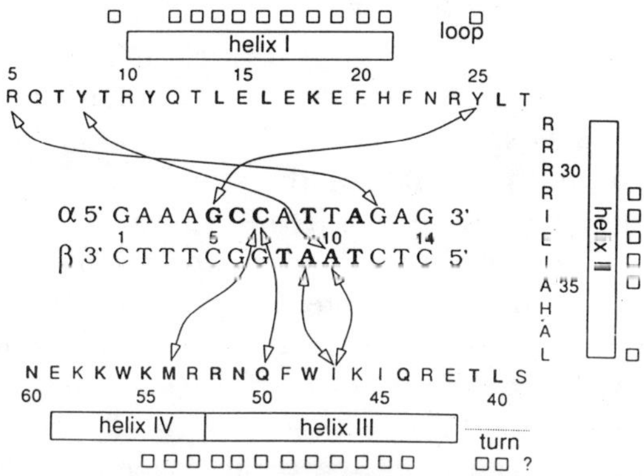

Fig. 3. Survey of the experimental NMR data collected with the 1 : 1 complex formed between the uniformly ^{15}N-enriched *Antp*(*C39*→*S*) homeodomain and the *BS2* DNA *14-mer*. The sequence of the *14-mer* is given in the center. The two strands are arbitrarily denoted α and β. The *Antp*(*C39*→*S*) amino acid sequence from residues 5 to 60 is arranged clockwise around the *14-mer*. The secondary structure of the protein observed both in the free form and in the complex is indicated alongside the amino acid sequence. Squares along the polypeptide sequence identify the residues with slowed amide proton exchange, indicating intramolecular hydrogen bonding. Bold letters identify those amino acid residues or nucleotides for which the ^{1}H chemical shifts differ by more than $|0.2|$ ppm from those in the free forms. Finally, the arrows indicate intermolecular contacts evidenced by NOEs between the protein and the DNA, where some arrows correspond to multiple NOEs between different protons of the amino acid residue and the nucleotide (reproduced from ref. *20*; numerous additional intermolecular NOEs have since been identified using the ^{13}C-labeled protein).

manifested by nuclear Overhauser effects (NOE) in the complex, that were previously observed for the free compounds. Additional information came from proton exchange measurements and measurements of scalar coupling constants between amide protons and α-protons (*24*). Both the slowed amide proton exchange associated with hydrogen bonds in helical secondary structures, and the small amide proton—α proton coupling constants indicate that the helix-boundaries are identical in the DNA-bound *Antp*(*C39→S*) homeodomain and in the free protein (*17*, *20*). The chemical shifts and the NOEs observed in the protein-bound 14-mer are typical for those of B-DNA (*24*).

Protein-DNA contacts in the 1 : 1 complex were directly identified from the *intermolecular* NOEs indicated by arrows in Fig. 3. With one exception the observed contact sites are all in the major groove of the DNA, the exception being the side chain of Arg 5 interacting with a sugar proton in the minor groove. Specific contacts (*i.e.*, contacts manifested by NOEs with base protons) of the homeodomain with the DNA were observed for the amino acid residues Ile 47, Gln 50, and Met 54. It should be emphasized that the relatively small number of ^1H-^1H NOEs in Fig. 3 enforce a much larger number of close intermolecular protein-DNA contacts (see below). Some of these additional contacts have since been identified in a DNA complex with the ^{13}C labeled *Antp*(*C39→S*) homeodomain (unpublished data), others are not directly observed by NMR because they do not involve hydrogen atoms.

The data of Fig. 3 were used as input for an ellipsoid algorithm calculation (*40*, *41*) to dock the protein against the DNA *14-mer*. In this initial structure determination a straight B-DNA conformation was maintained for the *14-mer*. For the *Antp*(*C39→S*) homeodomain we used the solution structure of the free protein with all amino acid residues replaced by alanine, because the spatial arrangement of the amino acid side chains in the protein-DNA interface could not be directly investigated using the ^{15}N-labeled protein. Although the actual steric repulsions are not fully accounted for with the use of this simplified protein structure, a unique orientation of the homeodomain relative to the DNA duplex was defined by the intermolecular NOEs (*20*). The structure of the complex is shown in Fig. 4, where the NOE distance constraints used in the docking calculations are indicated with bold lines. All these NOEs involve side chain protons of the polypeptide chain. The NOEs from Tyr 8 to the DNA β-strand (see Fig. 3 for the notation used) and from Tyr 25 to the DNA α-strand contact the backbone of the DNA on opposite sides of the major groove. Other NOE contacts are between Ile 47 and

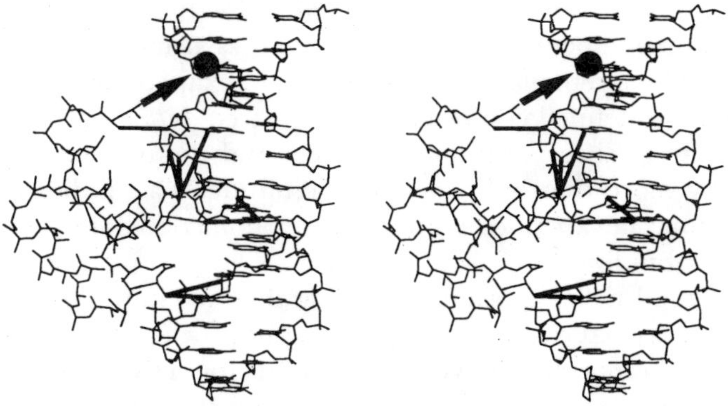

Fig. 4. Stereo view of the NMR structure of the *Antp(C39→S)* homeodomain-DNA complex. The complete heavy-atom structure of the DNA *14-mer* is shown. For the homeodomain only the polypeptide backbone and the β-carbon atoms of the residues 7–54 have been drawn. The NOE distance constraints used for the docking are drawn with bold lines. Note that these constraints connect in all instances to the β-carbons on the protein, even though in reality most of the NOEs have been observed with more peripheral side chain protons. An arrow points to a sphere in the minor groove of the DNA which identifies the location where NOEs with the side chain of Arg 5 (not drawn) were identified.

both the DNA backbone and a base proton inside the major groove, and from Gln 50 and Met 54 to a base proton inside the major groove. A sphere identifies an additional contact identified by a NOE from the side chain of Arg 5 (not shown in Fig. 4) to H1′ of G 12 in the minor groove. The drawing of Fig. 4 shows that the NOE contacts are spread over a large area of the molecular surface of the protein as well as the DNA. Outstanding features of the complex are the position of the elongated helix III in the major groove, which is determined by NOEs from three subsequent windings of this α-helix to base protons of the DNA, and the contacts between the minor groove of the DNA and the N-terminal region of the polypeptide chain, which is flexibly disordered in the free protein (*16*, *18*). With regard to earlier genetic and biochemical studies (*42*, *43*), it is worth noting that Gln 50 in position 9 of helix III does indeed show a NOE with C 7 of the DNA α-strand. A further noteworthy feature is the antiparallel arrangement of the polypeptide segment comprising residues Tyr 25, Leu 26, and Thr 27 with respect to the backbone of the DNA α-strand near A 4 and G 5.

3. *Biochemical Studies of Homeodomain-DNA Contacts*

The structure of the homeodomain-DNA complex as shown in Fig. 4 is in agreement with a variety of biochemical methylation and ethylation interference experiments (*38*) and with studies using mutant homeodomains (*44*). Thus, removal of the N-terminal hexapeptide segment, which does not have a defined conformation in the free *Antp* homeodomain (*16*), alters the interference patterns extensively and the dissociation constant increases by more than an order of magnitude. This indicates that the amino-terminal hexapeptide segment of the homeodomain is essential for high affinity binding.

The second helical position following the turn in the helix-turn-helix motif plays a crucial role for DNA binding in procaryotic repressors (*39*), but it appears not to be very important for DNA binding of the homeodomain, as indicated by the observation that *in vivo* reintroduction of an (Arg43→Ala) mutant *ftz* gene into *ftz*-flies complements the *ftz* mutant phenotype and results in normal development (M. Müller, K. Tokunaga, and W. Gehring, unpublished data). This is rationalized by the fact that compared to procaryotic repressors, the turn of the helix-turn-helix motif is shifted away from the DNA in the homeodomain-DNA complex (see Fig. 4).

In vivo and *in vitro* studies showed that a single amino acid substitution at position 9 of helix III is sufficient to change its binding specificity (*42*, *43*). Further extended *in vitro* DNA binding studies (*38*, *44*) showed that this amino acid (Gln 50 in the *Antp* homeodomain) interacts with the nucleotides C 6 and C 7 of the *BS2* binding site, in agreement with the NOE observed between Gln 50 and a base proton of C 7 (Fig. 3).

4. Comparison of the Antp(C39→S)-DNA Complex with a Procaryotic Repressor-DNA Complex

To compare the *Antp*(*C39→S*)-DNA complex with a procaryotic system, we selected the *434* repressor-DNA crystal structure (*45*, *46*). A single N-terminal 69-residue domain of the *434* repressor and the ten basepairs forming the binding "half-site" for this monomeric subunit were selected for the comparison. The homeodomain-DNA complex was superimposed to this structure by optimal fit of the corresponding helix-turn-helix motifs, which are very similar, with a root mean square deviation of the backbone atoms of 0.9 Å (*16*). The following three significant structural differences were readily apparent. (i) When compared to the *434* repressor-DNA complex, the DNA in the homeodomain complex is shifted by about 7 Å along helix III away from the turn toward the helix IV. As a result of this shift there are direct DNA contacts with helix IV of the homeodomain. (ii) In the *434* repressor complex the helix preceding the turn in the helix-turn-helix motif spans the major groove, which is also in direct contact with the third helix. In contrast, in the homeodomain complex (Fig. 4) the helix preceding the turn is located furthest away from the DNA near the surface of the complex. This, however, does not exclude that there might be direct DNA contacts with long side chains of residues in this helix, *e.g.*, one or several of the arginines (see Fig. 3). (iii) In the *434* repressor complex there is no counterpart for the type of minor groove contacts seen in the DNA complex with the *Antp* homeodomain (Fig. 4). Otherwise, the principal contacts of the homeodomain with the DNA *14-mer* do have counterparts in the *434* repressor-DNA complex. In particular, the NOEs from Tyr 8 of the homeodomain to the backbone of the DNA β-strand are paralleled by interactions of the residues preceding the fourth helix of the *434* repressor with phosphate groups of the DNA, and the NOEs from Tyr 25 to the backbone of the DNA α-strand have a counterpart in hydrogen bonds from the first residues of the second helix in the *434* repressor to DNA phosphates (*45*).

SUMMARY

Many of the master control genes regulating embryonic development and cell differentiation in eukaryotic organisms contain a highly conserved DNA segment, the *homeobox*. The corresponding polypeptide, the *homeodomain*, consists of about 60 amino acid residues and binds sequence-specifically to DNA. *In vitro*, the homeodomain binds to the *BS2* recognition site as a monomer with a dissociation constant of about 10^{-9} Mol/l. To understand the molecular basis of DNA recognition in the homeodomain-DNA system, the structures of the free *Antp* homeodomain in aqueous solution and the complex of the *Antp* homeodomain with a 14-base pair DNA fragment were investigated by NMR. The NMR data showed that the protein and the DNA have similar conformations in the free form and in the complex. Intermolecular NOEs observed between the ^{15}N-labeled protein and the DNA enabled a unique docking of the protein on the DNA. In the complex, the second helix of the helix-turn-helix motif of the homeodomain lies in the major groove of the DNA, and the amino-terminal arm reaches into the minor groove. The global features of this initial structure determination of a homeodomain-DNA complex were confirmed by subsequent crystallographic studies of DNA complexes with the *engrailed* homeodomain and with *Mat* α2, as well as by refinements of the NMR structure of the *Antp* complex using a ^{13}C-labeled *Antp*(*C39*$\rightarrow$*S*) homeodomain (unpublished data). The overall binding mode for homeodomains shows significant differences to that of procaryotic repressor proteins, although these also use a helix-turn-helix motif for DNA binding. The structural data have been shown to be in line with a large body of biochemical data accumulated about homeodomain-DNA recognition.

REFERENCES

1. W. Bender, M.E. Akam, F. Karch, P.A. Beachy, M. Peifer, P. Spierer, E.B. Lewis, and D.S. Hogness: *Science* **221**, 23–29 (1983).
2. R.L. Garber, A. Kuroiwa, and W.J. Gehring: *EMBO J.* **2**, 2027–2036 (1983).
3. M.P. Scott, A.J. Weiner, T.J. Hazelrigg, B.A. Polisky, V. Pirrotta, V.F. Scalenghe, and T.C. Kaufman: *Cell* **35**, 763–776 (1983).
4. W. McGinnis, M.S. Levine, E. Hafen, A. Kuroiwa, and W.J. Gehring: *Nature* **308**, 428–433 (1984).
5. W. McGinnis, R.L. Garber, J. Wirz, A. Kuroiwa, and W.J. Gehring: *Cell* **37**, 403–408 (1984).
6. M.P. Scott and A.J. Weiner: *Proc. Natl. Acad. Sci. U.S.A.* **81**, 4115–4119 (1984).
7. A. Fjose, W.J. McGinnis, and W.J. Gehring: *Nature* **313**, 284–289 (1985).
8. A.E. Carrasco, W. McGinnis, W.J. Gehring, and E.M. De Robertis: *Cell* **37**, 409–414 (1984).
9. W. McGinnis, C.P. Hart, W.J. Gehring, and F.H. Ruddle: *Cell* **38**, 675–680 (1984).
10. M. Levine, G.M. Rubin, and R. Tjian: *Cell* **38**, 667–673 (1984).
11. E. Boncinelli, A. Simeone, A. La Volpe, A. Faiella, V. Fidanza, D. Acampora, and L. Scotto: *Cold Spring Harbor Symp. Quant. Biol.* **50**, 301–306 (1985).
12. J.C.W. Shepherd, W. McGinnis, A.E. Carrasco, E.M. De Robertis, and W.J. Gehring: *Nature* **310**, 70–71 (1984).
13. F.D.H. Ohlendorf, W.F. Anderson, and B.W. Matthew: *J. Mol. Evol.* **19**, 109–114 (1983).
14. A. Laughon and M.P. Scott: *Nature* **310**, 25–31 (1984).
15. R.G. Brennan and B.W. Matthews: *J. Biol. Chem.* **264**, 1903–1906 (1989).
16. Y.Q. Qian, M. Billeter, G. Otting, M. Müller, W.J. Gehring, and K. Wüthrich: *Cell* **59**, 573–580 (1989).
17. M. Billeter, Y.Q. Qian, G. Otting, M. Müller, W.J. Gehring, and K. Wüthrich: *J. Mol. Biol.* **214**, 183–197 (1990).
18. P. Güntert, Y.Q. Qian, G. Otting, M. Müller, W.J. Gehring, and K. Wüthrich: *J. Mol. Biol.* **217**, 531–540 (1991).
19. C.L. Phillips, A.K. Vershon, A.D. Johnson, and F.W. Dahlquist: *Genes Dev.* **5**, 764–772 (1991).
20. G. Otting, Y.Q. Qian, M. Billeter, M. Müller, M. Affolter, W.J. Gehring, and K. Wüthrich: *EMBO J.* **9**, 3085–3092 (1990).
21. C.R. Kissinger, B. Liu, E. Martin-Blanco, T.B. Kornberg, and C.O. Pabo: *Cell* **63**, 579–590 (1990).
22. C. Wolberger, A.K. Vershon, B. Liu, A.D. Johnson, and C.O. Pabo: *Cell* **67**, 517–528 (1991).

23. M. Müller, M. Affolter, W. Leupin, G. Otting, K. Wüthrich, and W.J. Gehring: *EMBO J.* **7**, 4299–4304 (1988).
24. K. Wüthrich: "NMR of Proteins and Nucleic Acids" Wiley, New York (1986).
25. M.P. Scott, J.W. Tamkun, and G.W. Hartzell III: *BBA Rev. Cancer* **989**, 25–49 (1989).
26. A.K. Aggarwal, D.W. Rodgers, M. Drottar, M. Ptashne, and S.C. Harrison: *Science* **242**, 899–907 (1988).
27. C.O. Pabo and M. Lewis: *Nature* **298**, 443–447 (1982).
28. C. Wolberger, Y. Dong, M. Ptashne, and S.C. Harrison: *Nature* **335**, 789–795 (1988).
29. R. Zhang, A. Joachimiak, C.L. Lawson, R.W. Schevitz, Z. Otwinowski, and P.B. Sigler: *Nature* **327**, 591–597 (1987).
30. D.B. McKay and T.A. Steitz: *Nature* **290**, 744–749 (1981).
31. E.R.P. Zuiderweg, R. Kaptein, and K. Wüthrich: *Proc. Natl. Acad. Sci. U.S.A.* **80**, 5837–5841 (1983).
32. E.R.P. Zuiderweg, M. Billeter, R. Boelens, R.M. Scheek, K. Wüthrich, and R. Kaptein: *FEBS Lett.* **174**, 243–247 (1984).
33. R. Kaptein, E.R.P. Zuiderweg, R.M. Scheek, R. Boelens, and W.F. van Gunsteren: *J. Mol. Biol.* **182**, 179–182 (1985).
34. C. Desplan, J. Theis, and P.H. O'Farrell: *Nature* **318**, 630–635 (1985).
35. C. Desplan, J. Theis, and P.H. O'Farrell: *Cell* **54**, 1081–1090 (1988).
36. M. Affolter, A. Schier, and W.J. Gehring: *Curr. Opin. Cell Biol.* **2**, 485–495 (1990).
37. P.A. Beachy, M.A. Krasnow, E.R. Gavis, and D.S. Hogness: *Cell* **55**, 1069–1081 (1988).
38. M. Affolter, A. Percival-Smith, M. Müller, W. Leupin, and W.J. Gehring: *Proc. Natl. Acad. Sci. U.S.A.* **87**, 4093–4097 (1990).
39. M. Ptashne: "A Genetic Switch" Cell Press & Blackwell Sci. Publications (1986).
40. M. Billeter, T.F. Havel, and K. Wüthrich: *J. Comp. Chem.* **8**, 132–141 (1987).
41. M. Billeter, T.F. Havel, and I.D. Kuntz: *Biopolymers* **26**, 777–793 (1987).
42. S.D. Hanes and R. Brent: *Cell* **57**, 1275–1283 (1989).
43. J. Treisman, P. Gonczy, M. Vashishtha, E. Harris, and C. Desplan: *Cell* **59**, 553–562 (1989).
44. A. Percival-Smith, M. Müller, M. Affolter, and W.J. Gehring: *EMBO J.* **9**, 3967–3974 (1990).
45. A.K. Aggarwal, D.W. Rodgers, M. Drottar, M. Ptashne, and S.C. Harrison: *Science* **242**, 899–907(1988).
46. C.O. Pabo, A.K. Aggarwal, S.R. Jordan, L.J. Beamer, U.R. Obeysekare, and S.C. Harrison: *Science* **247**, 1210–1213 (1990).

Molecular Structure and Life (Ed. by Y. Kyogoku and Y. Nishimura), pp. 129–140, Japan Sci. Soc. Press, Tokyo/CRC Press, Boca Raton (1992)

Conformational Changes of Proteins and DNA Induced by Their Interactions

Masahiro SHIRAKAWA,*1 Hiroshi MATSUO*1, Yasuharu SERIKAWA,*1 Nobuaki SUZUKI*1, Chojiro KOJIMA,*1 Tadayasu OHKUBO,*2 and Yoshimasa KYOGOKU*1

*Institute for Protein Research, Osaka University**1 *Protein Engineering Research Institute**2

Recent structural studies of interactions between transcription factors and DNA have revealed that some protein-DNA interactions induce substantial conformational changes of protein or/and DNA. Particularly, many proteins have been demonstrated to bend DNA, these are *Escherichia coli* cAMP receptor protein (CRP) (*1*), IHF (integration host factor) (*2*), *E.coli* lexA protein (*3*), phage 434 repressor (*4*), phage 434 cro protein (*5*), NF-κB (*6*), JUN homodimer and JUN/FOS heterodimer (*7*).

In addition to their conformational change of DNA, some proteins were shown to induce such changes upon binding to DNA. For instance, when certain basic-region/leucine zipper (bZIP) proteins, such as yeast GCN4 (*8*) and JUN/FOS (*9*), are bound to a specific binding site in gene DNA, there is formation and stabilization of their helical structure.

We have examined three proteins to learn whether or not they have the ability to induce conformational change of DNA, using NMR, UV-CD (ultraviolet circular dichroism) and/or gel-electrophoresis. The proteins are cro protein from bacteriophage λ (*10*), the DNA binding

*1 Suita, Osaka 565, Japan
*2 Suita, Osaka 565, Japan

domain of pap1 protein from *Saccharomyces pombe* (*11*) and the DNA binding domain of yeast GAL4 (*12*) from *Saccharomyces cerevisiae*.

I. CONFORMATIONAL CHANGE OF λ-Cro PROTEIN INDUCED BY SPECIFIC INTERACTION WITH O_R3 DNA (*13*)

λ-Cro protein is a repressor protein which binds to the right (O_R) and left (O_L) operator regions of the λ phage genome competitively with the cI repressor. It consists of 66 amino acids and acts as a dimer. Cro protein most strongly binds to an operator region called O_R3, which consists of 17 base pairs (*14*).

The ^{1}H-^{15}N spectrum of the uniformly ^{15}N-labeled cro protein is shown in Fig. 1. Assignments of the cross peaks and identification of secondary structure were previously reported (*15*). On the addition of an equimolar amount of λ-O_R3 17-mer double-stranded DNA, which is the specific binding site of the cro protein, to a solution of the cro protein,

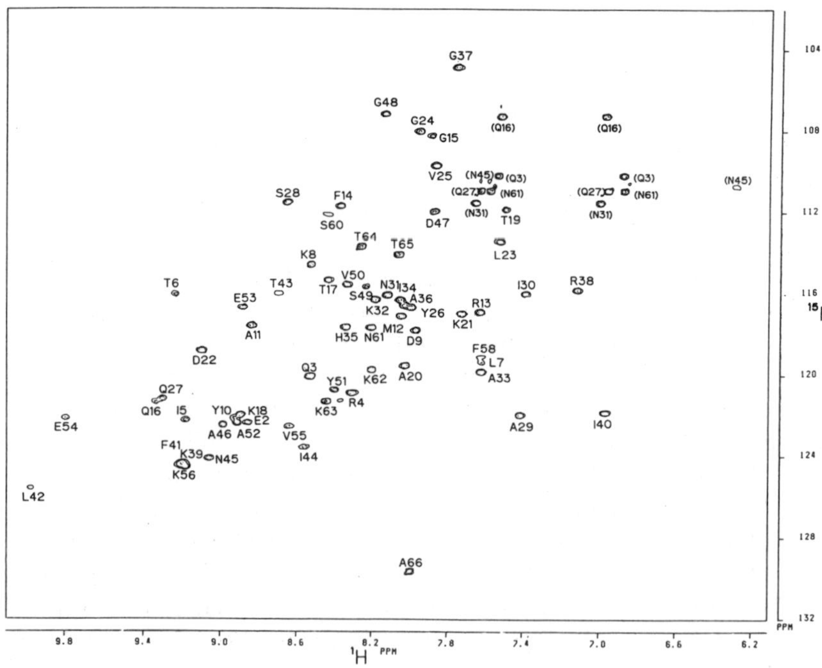

Fig. 1. ^{1}H-^{15}N SQC spectrum of the λ-cro protein (*15*). The signal assignments are indicated by amino acid residue numbers. Numbers in parentheses show the side chain amide proton signals.

some cross peaks shifted markedly on both the ^{1}H axis and ^{15}N axis (Fig. 2). In the titration most of the cross peaks were broadened, probably due to the increment of their molecular weight upon complexation, in some the broadening was remarkable. The chemical shifts of amide proton resonances and amide nitrogen-15 resonances are plotted in Fig. 2a and 2c.

To determine whether or not the induced shifts of the amide proton and amide nitrogen-15 resonances are due to base sequence specific interaction with DNA, $^{1}H/^{15}N$ resonances of the cro protein were also examined for the complex with 10-mer double-stranded DNA, which contains no apparent base sequence homology with λ-O_R3. The chemical shifts of the amide proton resonances and amide nitrogen-15 resonances of the complex with the 10-bp DNA are plotted in Fig. 2b and 2d.

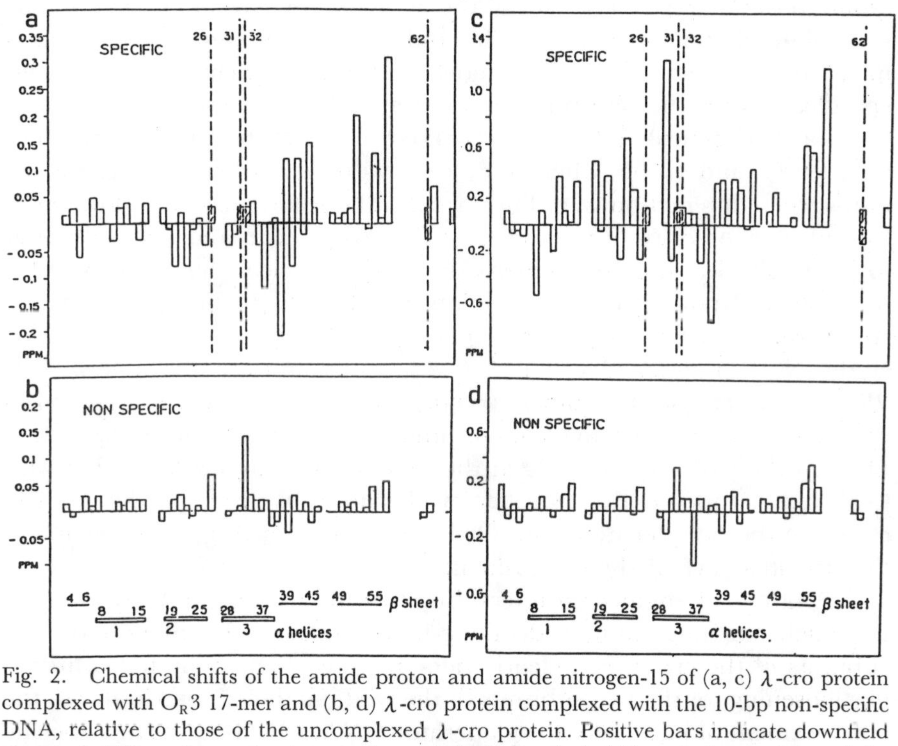

Fig. 2. Chemical shifts of the amide proton and amide nitrogen-15 of (a, c) λ-cro protein complexed with O_R3 17-mer and (b, d) λ-cro protein complexed with the 10-bp non-specific DNA, relative to those of the uncomplexed λ-cro protein. Positive bars indicate downfield chemical shifts and negative bars indicate upfield chemical shifts, relative to those of the uncomplexed λ-cro protein (*15*). Broken lines indicate residues of which signals showed such large chemical shifts or so much broadening that the assignments in the complex were not obtained.

As each amide proton and amide ^{15}N signal has been assigned to each residue completely, it is clearly apparent which residues are affected by the interaction with O_R3 and/or the 10-bp non-specific DNA fragment. The induced chemical shifts are significantly larger than those observed in the non-specific interaction with the 10-bp DNA. ^1H and ^{15}N chemical shifts of the α_3 helix (residues 28–36) are remarkable in both specific and non-specific interaction. This is consistent with our previous photo-CIDNP (chemically induced dynamic nuclear polarization) study (*16*), and the crystal structure of cro-O_R3 complex by Brennan and co-workers (*17*), both of which showed that the α_3 helix is involved in the direct interaction with DNA. The chemical shift changes observed for the residues in the α_3 helix in the specific interaction were significantly larger than those observed in the non-specific interaction. Some resonances had too large a broadening, and were therefore difficult to observe. In addition to $^1H/^{15}N$ resonances of the α_3 helix, some resonances of the α_2 helix showed shifts in both specific and non-specific interactions. This is in good agreement with the model that the α_2 helix interacts non-specifically with the phosphate backbone of DNA.

Large chemical shifts were observed in the β-sheet region (residues 4–6, 39–45, and 49–55) when complexed with O_R3, whereas those for the non-specific interaction were fairly small. The β-sheet region is located at the subunit interface of the dimer of the cro protein, and is not thought to be involved in the direct interaction. The chemical shifts observed for the β-sheet upon specific interaction are thought to be caused by the structural change in the β-sheet region. Since our previous photo-CIDNP study, NOE (nuclear Overhauser effect) study and mutant analysis (*16, 19, 20*) showed that the dimer structure of the cro protein is loosened in the region of inter-subunit contact and a conformational change takes place in the hydrophobic core at the time of the complex formation with O_R3, it is supposed that the chemical shifts observed for the β-sheet region in the complexation with O_R3 are accompanied by a change in the dimeric structure of the cro protein.

Analysis of engineered intersubunit disulfide-bonded λ-cro protein, in which the intersubunit disulfide-bond restricts the motion of two subunits of the cro dimer, clearly demonstrated that the induced fitting, like loosening, of the two subunits of the wild cro dimer contributes to the enhancement of its affinity to its operator DNA, which results in a specific interaction between cro and O_R3 (*21*).

Analysis of the imino proton resonances of O_R3 17-mer demonstrated that a conformational change of DNA also takes place upon complexation with the cro protein (*10*).

Taking all the previous experimental data together, we can infer that molecular recognition of cro protein-O_R3 is achieved by facilitation of the induced fitting of both the protein and DNA, which results in tight binding and specific interactions between amino acids and bases.

II. DNA BEND INDUCED BY BINDING OF A LEUCINE ZIPPER PROTEIN, PAP-1

The *pap1*$^+$ gene codes a protein of 544 amino acids (designated pap1 protein), of which the deduced amino acid sequence contains a "leucine zipper motif" (*22*). We have recently shown that pap1 protein binds to two different sequences of the upstream region of a fission yeast gene p25 (called AP-1 site and PLD site, respectively; these sequences are shown in Fig. 3) in a sequence specific manner, as well as to the AP-1 site in SV-40 enhancer which is a binding site of JUN/AP-1 (*11*, *23*).

The interaction of the 70 amino acid DNA binding domain of pap1 protein (designated PAP1(70)) and its binding sequences was studied by UV-CD. We prepared double-stranded oligonucleotides, which contain either sequence of the two binding sites found in the upstream region of p25, and O_R3 of λ-phage, respectively. The positive circular dichroism band of AP-1 and PLD sites showed remarkable increment when titrated with PAP1(70). On the other hand, the observed increment of the CD band of λ-O_R3, which does not bind to pap1 protein in a sequence specific manner (*11*), was relatively small upon the titration with PAP1-(70) (Fig. 4).

AP-1 site	CTAGAGAATAGGATTAGTCAGAAATGG TCTTATCCTAATCAGTCTTTACCAGCT
PLD site	CTAGAGGTAGTTTACGTAATCTTGGTAG TCCATCAAATGCATTAGAACCATCAGCT
λ-O_R3	TATCACCGCAAGGGATA ATAGTGGCGTTCCCTAT

Fig. 3. Nucleotide sequences of the three DNA duplexes used for the DNA binding property of the DNA binding domain of pap1 protein (PAP1(70)) (*12*, *23*). AP-1 consensus binding site and the semipalindromic 14-bp sequence are boxed.

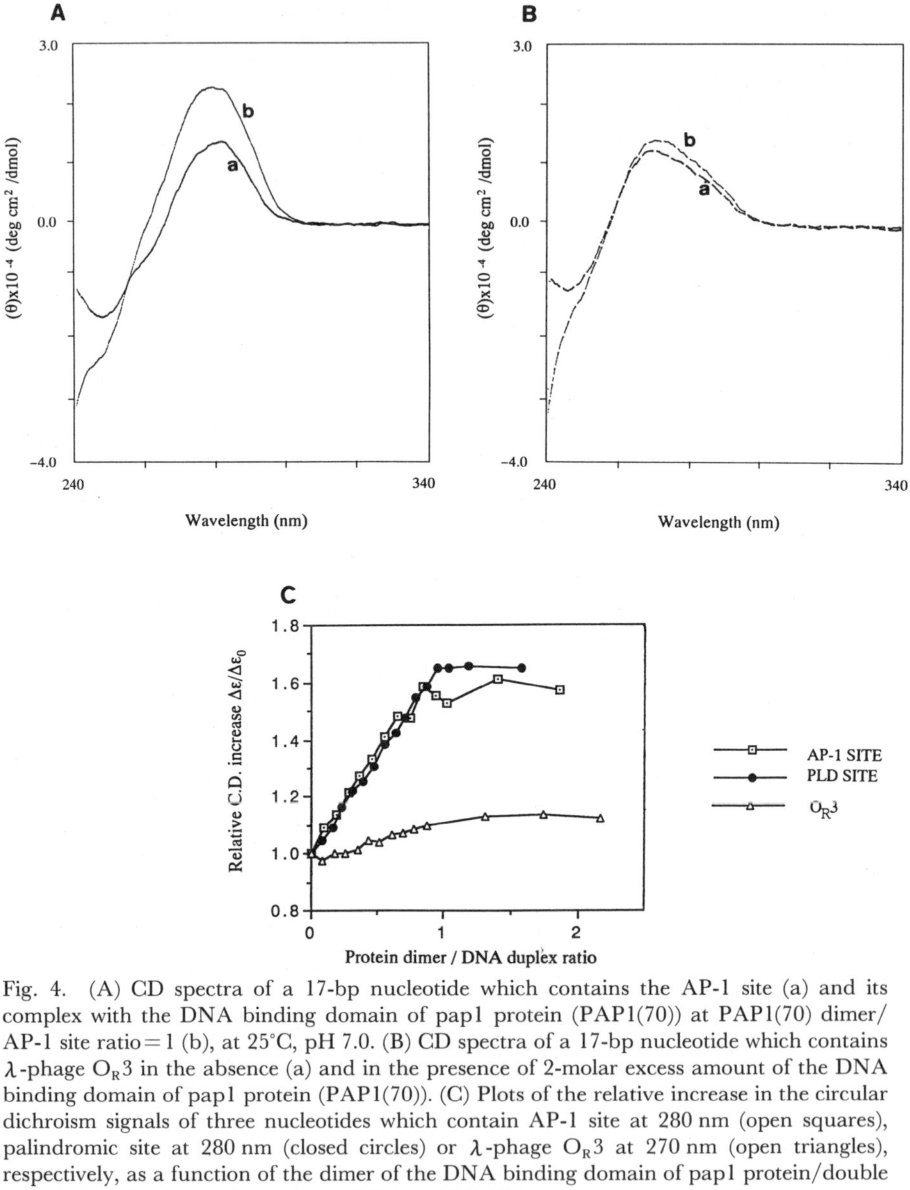

Fig. 4. (A) CD spectra of a 17-bp nucleotide which contains the AP-1 site (a) and its complex with the DNA binding domain of pap1 protein (PAP1(70)) at PAP1(70) dimer/AP-1 site ratio = 1 (b), at 25°C, pH 7.0. (B) CD spectra of a 17-bp nucleotide which contains λ-phage O_R3 in the absence (a) and in the presence of 2-molar excess amount of the DNA binding domain of pap1 protein (PAP1(70)). (C) Plots of the relative increase in the circular dichroism signals of three nucleotides which contain AP-1 site at 280 nm (open squares), palindromic site at 280 nm (closed circles) or λ-phage O_R3 at 270 nm (open triangles), respectively, as a function of the dimer of the DNA binding domain of pap1 protein/double stranded nucleotide molar ratio, at 25°C, pH 7.0 (*11*).

A similar phenomenon was observed for the interaction of lexA repressor with poly[d(A-T)] by Schnarr and co-workers (*24*), suggesting that it induces a similar conformational change of DNA. Using the gel electrophoresis analysis of Lloubes *et al.* (*3*), it was shown that DNA was bent in the lexA-DNA system.

DNA bending induced upon complexation with PAP1(70) was analyzed by circular permutation mobility assay (*25, 26*). AP-1 site and PLD site sequences were independently cloned into the same position of plasmid pBEND2 (*27*), and several restriction fragments from each plasmid were complexed with PAP1(70). These fragments were of identical length, containing either AP-1 site or PLD site at different positions relative to the ends of the fragments. Figure 5 shows comparative gel-electrophoretic data for circularly permuted variants of DNA fragments containing AP-1 site and PLD site complexed with PAP1(70). From the result, it is inferred that upon binding to either sequence PAP1(70) is able to induce a substantial bending of the DNA. Comparing lane 6 and lane 12, it seems that DNA in the AP-1 site complex is slightly more bent than that in the PLD site complex. The comparative gel-electrophoretic data for circularly permuted variants of uncomplexed DNA fragments containing AP-1 site or PLD site did not show any bending scheme (data not shown). However, phase-sensitive DNA bend assay (*7, 28*) is necessary to prove that the conformational change of the DNA in the PAP1(70)-DNA complex is bending.

The role of DNA bending found in transcription regulation systems is not yet clear. A model has been proposed to explain the role of CRP induced DNA bending; the induced bend facilitates protein-DNA or protein-protein contacts that would be sterically difficult on a straight promoter DNA (*26*). It can be inferred that pap1 protein induced bend of the upstream region of p25 at two sites located 52 bp apart might facilitate protein-DNA and/or protein-protein interaction which regulate the transcription initiation of p25 (*11*). Recently, Toda *et al.* (*23*) found that both of the two sites in the upstream region of p25 are necessary for pap1 protein dependent *trans*-activation.

III. INTERACTION OF THE DNA BINDING DOMAIN OF YEAST GAL4 AND DNA

GAL4, a transcriptional activator protein that binds specifically to 17 base-pair DNA sequences called UAS_G (*29*), is responsible for induction of the genes required for galactose catabolism in *S. cerevisiae* (*30*). We

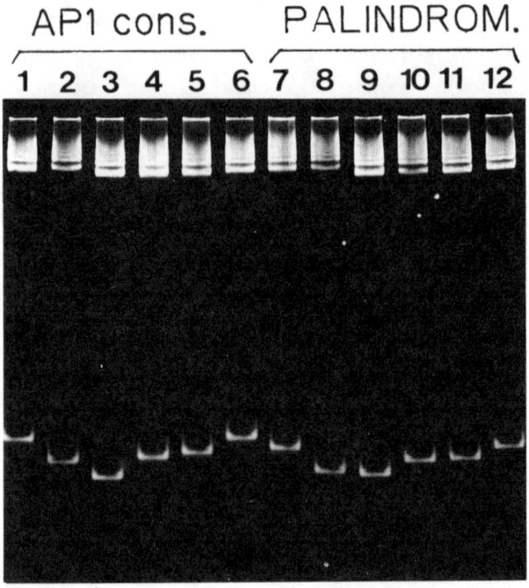

Fig. 5. 10% polyacrylamide gel electrophoretic mobility shift analysis of the DNA binding domain of pap1 protein bound to circularly permuted probes which contain the AP-1 site (lanes 1–6) or the palindromic site (lanes 7–12) (*11*). The DNA binding domain of pap1 protein was incubated with the NruI fragments (lanes 1 and 7), the BamHI fragments (lanes 2 and 8), the MluI fragments (lanes 3 and 9), the BglII fragments (lanes 4 and 10), the NheI fragments (lanes 5 and 11) and the EcoRV fragments (lanes 6 and 12) from pBEND2 (*28*) derivatives which contain either the AP-1 site or the palindromic site at the XbaI-SalI site.

earlier showed that GAL4 fragments of 96 and 110 amino-terminal residues possess the specific binding activity to UAS_G, and that Zn(II) is necessary for the conformation of the DNA binding domain as well as for the specific DNA binding (*31*).

The interaction of the 110 residue DNA binding domain of GAL4 and its binding site, UAS_G, was studied by monitoring imino proton resonances of oligodeoxynucleotide. We synthesized 17-mer deoxyribonucleotide duplex chemically; this is one of the UAS_G sequences in yeast *gal1-gal10* promoter (the sequence is shown in Fig. 6A). The imino proton region of the ^{1}H NMR spectra of the 17-bp deoxyribonucleotide duplex in aqueous solution at 20°C is shown at the top in Fig. 6B. All of the imino proton resonances were assigned to individual base-pairs by means of sequential NOE tracing (*12*). On addition of the 110-amino terminal residues of GAL4, which possess the sequence-specific DNA binding activity (*31*), all of the imino proton signals became broader. The spectrum at the 1 : 1 ratio (dimer of the DNA binding domain of GAL4/DNA

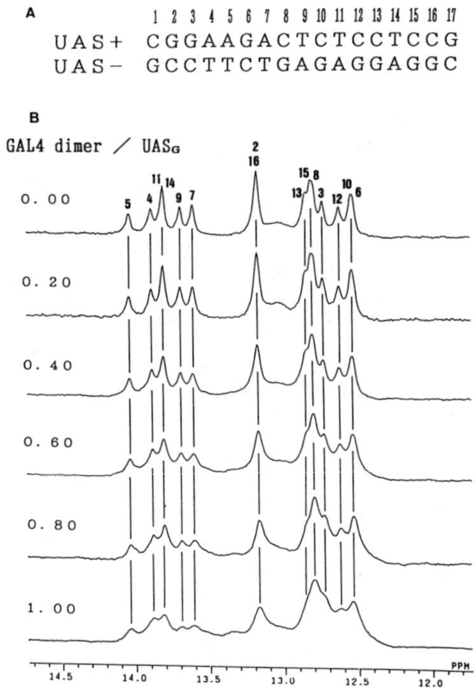

Fig. 6. (A) The nucleotide sequence of UAS_G 17-mer. This sequence is one of the four GAL4 binding sites of *gal1-gal10* gene from *S. cerevisiae.* (B) The imino proton resonances of UAS_G 17 mer complexed with the DNA binding domain of GAL4 [GAL4(110)] at different GAL4(110) dimer/UAS_G ratios, at 20°C, pH 7.0 (*12*).

duplex) is given at the bottom in Fig. 6B. The broadening of most of the resonance lines may be explained by the increase in molecular weight of the complex. However, plots of the chemical shifts of the imino proton signals against the ratios to the added protein showed that only a small shift could be detected with the addition of an equimolar amount of the DNA binding domain of GAL4 (Fig. 6B). The binding of the 110-amino terminal residues of GAL4 and UAS_G 17-mer was analyzed earlier by gel-retardation assay (*31*).

The analysis of the imino proton resonances of O_R3 17-mer duplex DNA revealed large chemical shifts for some imino proton signals during the titration with λ-cro protein, indicating that a conformational change of DNA takes place upon complexation with λ-cro protein (*10*). A similar phenomenon was observed for pap1 protein/DNA interaction, where the imino proton resonances of 15-mer DNA duplex which contains the sequence of AP-1 site in SV40 enhancer (*32*) showed large induced chemical shifts. Conformational changes of DNA for these two protein-DNA interactions were also shown by experiments using gel electrophoresis and UV-CD by Shirakawa and co-workers (*11*), Lyubchenko and co-workers (*33*), and by Kirpichnikov and co-workers (*34*).

The imino proton signals of DNA are convenient markers for conformational DNA changes induced by the binding of proteins, because the chemical shifts of the imino protons within the double-stranded DNA are highly sensitive to the ring-current effects of the neighboring base-pairs and thus are strongly influenced by the local geometry and chemical environment. Therefore, the fact that the imino proton resonances of UAS_G showed only small shifts upon complexation with the DNA binding domain of GAL4 indicates that no or only a slight conformational change of the oligonucleotide takes place with binding of the protein. Similar evidence was also shown by UV-CD analysis of UAS_G, where only slight change was observed for the positive circular dichroism band (250–300 nm) upon complexation with the DNA binding domain of GAL4 (Hansen *et al.* (*18*)).

SUMMARY

Recent physicochemical studies of interactions between transcription factors and DNA have revealed that some protein-DNA interactions induce substantial conformational changes of protein or/and DNA. We examined three proteins to learn whether or not they have the ability to induce conformational change of DNA: cro protein from bacteriophage λ, the DNA binding domain of pap1 protein from *S. pombe* and the DNA binding domain of GAL4 from *S. cerevisiae*.

The conformational change of the operator DNA fragment induced by binding of cro protein was shown by analyzing the imino proton resonances of the nucleotide; it was also studied by ligation assay, UV-CD and crystallographic study by other investigators. Bending type conformational changes were also inferred to be induced for the two DNA fragments upon complexation with fission yeast pap1 protein, and were demonstrated by UV-CD and circular permutation mobility assay.

However, contrary to the two cases described above, the study of the imino proton resonances of UAS_G, a specific binding site of yeast GAL4, showed that little or no meaningful conformational change on the DNA fragment was induced when the base sequence specific complex was formed with the DNA binding domain of GAL4.

In the interaction of cro protein-DNA, a conformational change of the cro protein was also observed which was analyzed by stable isotope aided NMR and photo CIDNP. Analysis of engineered intersubunit disulfide-bonded λ-cro protein, in which intersubunit disulfide-bond restricts the motion of two subunits of the cro dimer, clearly demonstrated that the induced fitting, like loosening, of the two subunits of the wild cro dimer contributes to the enhancement of its affinity to its operator DNA, which results in a specific interaction between cro and O_R3.

REFERENCES

1. H.-N. Liu-Johnson, M.R. Gartenberg, and D.M. Crothers: *Cell* **47**, 995-1005 (1986).
2. C.A. Robertson and H.A. Nash: *J. Biol. Chem.* **263**, 3554-3557 (1988).
3. R. Lloubes, M. Granger-Schnarr, C. Lazdunski, and M. Schnarr: *J. Mol. Biol.* **204**, 1049-1054 (1988).
4. A.K. Aggarwal, D. Rodgers, M. Drottar, M. Ptashne, and S.C. Harrison: *Science* **242**, 899-907 (1988).
5. A. Mondragon and S.C. Harrison: *J. Mol. Biol.* **219**, 321-334 (1991).
6. R. Schreck, H. Zorbas, E.-L. Winnacker, and P.A. Baeuerle: *Nucl. Acids Res.* **18**, 6497-6502 (1990).
7. T.K. Kerppola and T. Curran: *Cell* **66**, 317-326 (1991).
8. M. Weiss, T. Ellenberger, C.R. Wobbe, J.P. Lee, S.C. Harrison, and K. Struhl: *Nature* **347**, 575-578 (1990).
9. L. Patel, C. Abate, and T. Curran: *Nature* **347**, 572-575 (1990).
10. S.J. Lee, M. Shirakawa, H. Akutsu, Y. Kyogoku, M. Shiraishi, K. Kitano, M. Shin, E. Ohtsuka, and M. Ikehara: *EMBO J.* *6*, 1129-1135 (1987).
11. M. Shirakawa, T. Toda, N. Suzuki, M. Shimanuki, M. Yanagida, and Y. Kyogoku: in preparation.
12. C. Kojima, M. Shirakawa, Y. Serikawa, and Y. Kyogoku: in preparation.
13. M. Shirakawa, H. Matsuo, T. Ohkubo, and Y. Kyogoku: in preparation.
14. M. Ptashne: "A Genetic Switch", Cell Press and Blackwell Scientific, Cambridge (1986).
15. H. Matsuo, M. Shirakawa, T. Ohkubo, T. Yamazaki, and Y. Kyogoku: *J. Biomol. NMR* **1**, 191-204 (1991).
16. M. Shirakawa, S.J. Lee, H. Akutsu, Y. Kyogoku, K. Kitano, M. Shin, E. Ohtsuka, and M. Ikehara: *FEBS Lett.* **181**, 286-290 (1985a).
17. R.G. Brennan, S.L. Roderick, Y. Takeda, and B.W. Matthews: *Proc. Natl. Acad. Sci. U.S.A.* **87**, 8165-8169 (1990).
18. A. Hansen, M.V. Hoy, and T. Kodadek: *Biochem. Biophys. Res. Commun.* **175**, 492-499, (1992).
19. M. Shirakawa, S.J. Lee, K. Yamamoto, M. Takimoto, H. Akutsu, and Y. Kyogoku: In "Structure and Expressions, from Proteins to Ribosomes" (Ed. by R.H. Sarma and H.H. Sarma), Vol. 1, pp. 167-179, Adenine Press (1987).
20. M. Shirakawa: Doctoral thesis, School of Science, Osaka University (1988).

21. M. Shirakawa, H. Matsuo, and Y. Kyogoku: *Protein Engineering* **4**, 545–552 (1991).
22. T. Toda, M. Shimanuki, and M. Yanagida: *Genes Dev.* **5**, 60–73 (1991).
23. T. Toda, M. Shimanuki, Y. Saka, H. Yamano, Y. Adachi, M. Shirakawa, Y. Kyogoku, and M. Yanagida: *Mol. Cell. Biol.*, in press.
24. M. Schnarr and M. Daune: *FEBS Lett.* **171**, 207–210 (1984).
25. H.-M. Wu and D.M. Crothers: *Nature* **308**, 509–513 (1984).
26. H.-N. Liu-Johnson, M.R. Gartenberg, and D.M. Crothers: *Cell* **47**, 995–1005 (1986).
27. J. Kim, C. Zwieb, C. Wu, and S. Adhya: *Gene* **85**, 15–23 (1989).
28. S.S. Zindel and D.M. Crothers: *Nature* **328**, 178–181 (1987).
29. E. Giniger, M.S. Varnum, and M. Ptashne: *Cell* **40**, 767–774 (1985).
30. Y. Oshima: In "Molecular Biology of the Yeast Saccharomyces" (Ed. by J. Strathem, E. Jones, and J.K. Broach), Vol. 1, pp. 159–180, Cold Spring Harbor Laboratory Press, Cold Spring Harbor, New York (1981).
31. Y. Serikawa, M. Shirakawa, H. Matsuo, and Y. Kyogoku: *Protein Engineering* **3**, 267–272 (1990).
32. C. Kojima, M. Shirakawa, and Y. Kyogoku: unpublished results.
33. Y. Lyubchenko, L. Shlyakhtenko, B. Chernov, and R.E. Harrington: *Proc. Natl. Acad. Sci. U.S.A.* **88**, 5331–5334 (1991).
34. M.P. Kirpichnikov, A.P. Yartzev, L.E. Minchenkova, B.K. Chernov, and V.I. Ivanov: *J. Mol. Struct. Dynamics* **3**, 529–536 (1985).

Molecular Structure and Life (Ed. by Y. Kyogoku and Y. Nishimura), pp. 141–156, Japan Sci. Soc. Press, Tokyo/CRC Press, Boca Raton (1992)

Recognition of Double- and Single-stranded Substrates by Nucleases: Structural Studies of DNase I and P1 Nuclease

Dietrich SUCK

*European Molecular Biology Laboratory**

In recent years a steadily growing number of high resolution X-ray structures as well as some NMR studies of protein-nucleic acid complexes have given us some insight into the structural basis of the recognition process (*1*). In the complexes specific contacts are formed with the functional groups of the bases exposed in the grooves of double helical substrates enabling a "direct readout" of the base sequence. In addition, numerous, mainly electrostatic contacts involving the sugar-phosphate backbone have been found in all specific complexes studied so far. In most complexes deviations of varying degree from the standard helix geometry have been observed, confirming the now well established fact that DNA is an intrinsically flexible molecule, which shows considerable sequence-dependent variations of local helix parameters and which can easily be distorted by proteins. Thus specific recognition also seems to involve an "indirect readout" of the base sequence.

Sequence-dependent variations of local helix geometry are also recognized by DNase I, an endonuclease whose cutting rates vary strongly with the base sequence (*2*). We are studying the structural basis of this

* 6900 Heidelberg, Germany

sequence-dependence by solving the high resolution structures of DNase I-DNA complexes. So far, the structures of two octamer duplexes have been determined and will be discussed (*3–5*).

Relatively little is known yet at the structural level about the interactions of proteins with single-stranded DNA or RNA. One expects that in addition to the contacts to the functional groups of the bases and to the sugar-phosphate backbone, stacking interactions with the nucleobases should play an important role. Indeed, a common feature in complexes with single-stranded nucleic acids appears to be the stacking of exposed aromatic side chains with the nucleobases. We have solved the structure of the single-strand specific P1 nuclease by X-ray analysis (*6*) and are studying the binding of substrates by crystal soaking experiments.

I. DNase I

1. Properties of DNase I

Bovine pancreatic deoxyribonuclease I (DNase I) is a glycoprotein of molecular weight 30,400 and functions as a digestive enzyme by degrading preferentially double-stranded DNA through hydrolysis of the P-03′-bond (*7*, *8*).

The enzyme has been used as a biochemical probe in a number of experiments (*9*, *10*) and is now an established tool in the molecular biology laboratory. Although DNase I is neither sequence nor base specific it exhibits a strong sequence dependence (*11*, *2*). Cleavage probabilities are very low in regions of DNA consisting of runs of consecutive A's or T's and, to a lesser extent, are also lowered in G,C-rich regions compared to regions of mixed sequence. Probabilities at neighboring phosphodiester bonds may vary drastically, by up to a factor of 400. Thus it appears that DNase I senses sequence dependent structural variations in DNA. Individual helix parameters have been suggested to account for the observed cleavage pattern (*11–13*). Drew and Travers (*2*) first realized that the width of the minor groove is an important determinant of cleavage rates. However, given the rather complex cleavage pattern, we believe that only a combination of local and global helix parameters will fully account for the difference in cutting rates.

Although numerous studies have dealt with the properties and biochemistry of DNase I over the years (for reviews, see *7*, *8*), the detailed nature of the catalytic mechanism remained largely unclear. Early on, however, it was clear that divalent cations as well as H134 play an essential role in catalysis (*14*). Later on, NMR studies showed that

DNase I cleaves the P-03′-bond of the DNA substrate with inversion of the configuration at the phosphorous (*15*). This suggests that the hydrolysis occurs through direct nucleophilic attack of a water molecule in an in-line mechanism. The high resolution crystal structures of the enzyme and its complexes with oligonucleotide duplexes (*3–5*, *16–18*) and site-directed mutagenesis studies (*19*) suggest that two histidines (*H134*, *H252*) are directly involved in catalysis.

2. *Architecture of the DNase I Molecule*

The following description of the 3-D structure of DNase I is based on X-ray structure analysis and refinement at 2 Å resolution (*16*, *17*). The 260 amino acid residues of DNase I fold into a compact α,β-protein of approximate dimensions $45 \times 40 \times 35$ Å. The carbohydrate side chain attached *via* an N-glycosidic bond to N18 protrudes by about 15 Å from the surface of the globular molecule (Fig. 1). The central core of DNase I is formed by two tightly packed six-stranded β-sheets and is surrounded by three longer and five shorter α-helices and several loop regions. This arrangement gives rise to a four-layer sandwich-type structure. The extended hydrophobic core, comprising about 30% of all residues, is mainly responsible for the structural stability and rigidity of DNase I. Other factors contributing to the stability of the enzyme (besides numerous internal hydrogen bonds and several salt bridges), are two disulphide bridges, one of which (C173-C209) is essential for activity, and two

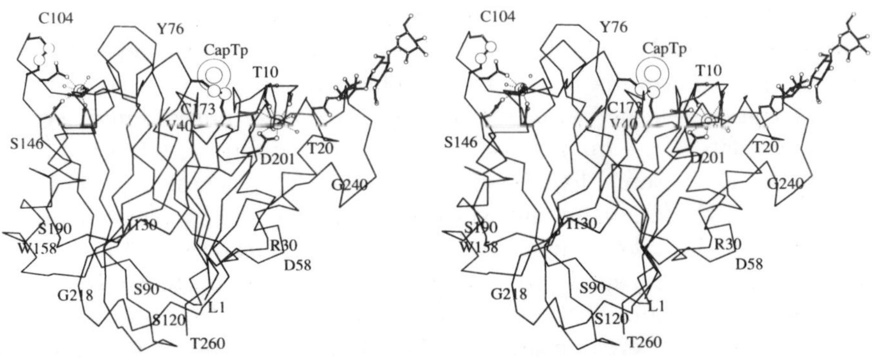

Fig. 1. C_α-backbone of DNase I. Open circles indicate the disulphide bridges connecting C101-C104 and C173-C209. The two structural Ca^{2+}-binding sites are indicated by small double circles and are shown together with their ligands. The position of the catalytic metal binding site determined by binding CapdTp in the crystal is marked by a large double circle. In the upper left corner the carbohydrate attached to N18 is shown.

Ca^{2+}-binding sites which stabilize loop regions in the molecule (*20*, *21*). A third low affinity Ca^{2+}-binding site exists at the active site but is occupied only in the presence of substrate, and was identified by soaking CapdTp in the native crystals (*16*).

3. The 3-D Structures of Two DNase I-DNA Complexes

As with other protein-DNA complexes (*22*, *23*), various length oligonucleotides were used in a trial and error approach for cocrystallization experiments. Well diffracting DNase I cocrystals were obtained with two self-complementary octanucleotides, $d(GCGATCGC)_2$ (Octa I) and $d(GGTATACC)_2$ (Octa V). The structures of the complexes were solved by molecular replacement using the known DNase I coordinates and refined to 2.0 Å (Octa I) and 2.3 Å (Octa V), respectively (*3*, *4*, 24).

A surprising result of the X-ray analysis of the DNase I-Octa I complex was the finding that, in spite of the presence of EDTA, the octamer duplex is nicked in one strand and a dinucleotide (pGpC) removed from the 3′ end before crystallization. In the crystal, the nicked duplexes dimerize across a twofold axis with base-pair formation between the single-stranded ends, producing a double helix of 14 base pairs which lacks one phosphate in each strand. The crystal contains discrete 14-mer units stacking with their 5′ terminal G's onto H121 residues of symmetry-related DNase I molecules. One of these quasi-continuous 14-mers bound to a DNase I molecule is shown in Fig. 2. The basic features of the complex are consistent with a model proposed earlier on the basis of the 3-D structure of the enzyme (*18*). The exposed loop around Y76 of DNase I binds tightly in the minor groove of a B-type DNA and positively charged or hydrophilic residues interact with several phosphates of both strands of the DNA. The contacts are restricted to one side of the double helix and extend over six base pairs. There are no direct contacts in the major groove, but the protein approaches the DNA backbone again some five base pairs away. The access to the major groove 5′ to the cleavage site will therefore be partially blocked, a fact which has to be taken into account when one interprets DNase I footprinting results. Except for a few side chains at the protein-DNA interface, no conformational change takes place in DNase I on binding of the oligonucleotide, whereas numerous distortions are induced in the DNA.

The interactions involved in complex formation are predominantly of electrostatic character, although van der Waals and stacking type contacts also occur (Fig. 2). Several hydrogen bonds of either backbone NH groups (E13, T14, S43) or side chains (R9, E39, N74, Y175, S206,

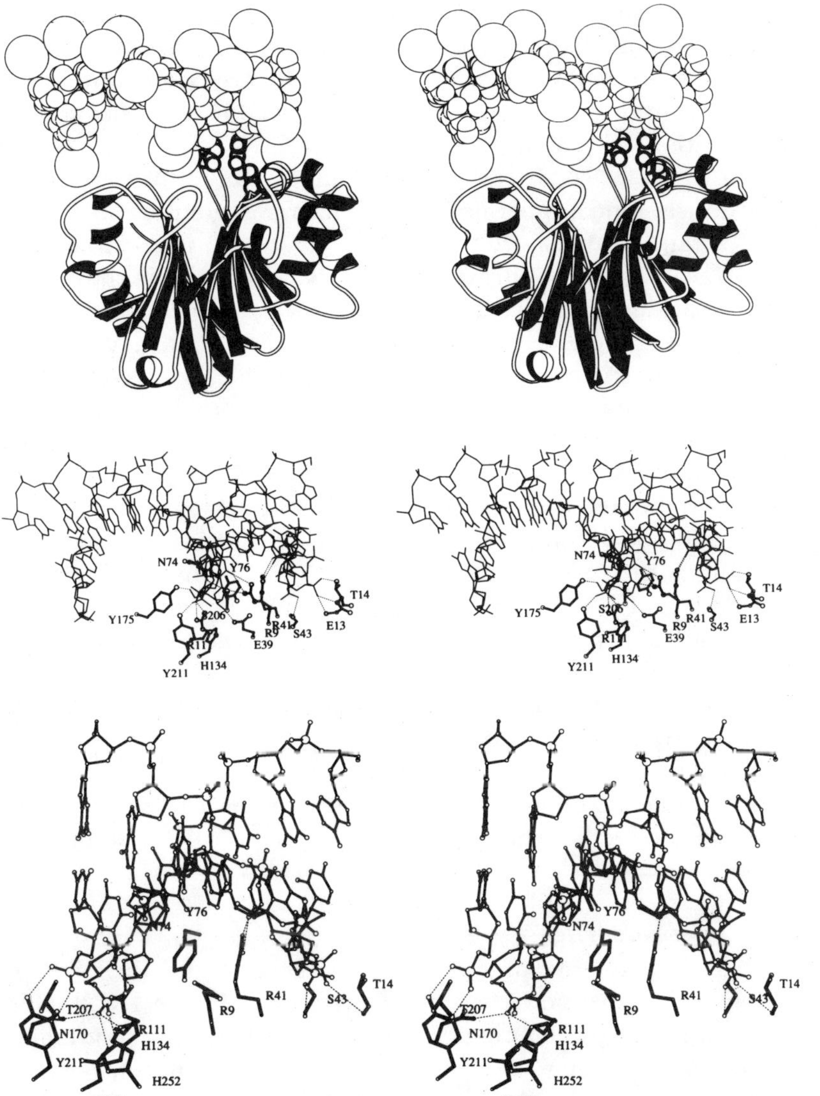

Fig. 2. 3D-structure of two DNase I-DNA complexes. Schematic stereo view of the DNase I-Octa I complex. A quasi-continuous 14mer drawn in space-filling mode is shown bound to DNase I. R41 and Y76 forming tight interactions in the minor groove of DNA are also represented in spacefilling mode (top). Stereo representation of the protein-DNA contacts in the DNase I-Octa I complex (middle). Protein-DNA contacts in the DNase I-Octa V complex (bottom).

Y211), and one salt bridge (R111) are formed with the sugar-phosphate backbones. Within the minor groove, R41 forms hydrogen bonds with the O2 atoms of two successive pyrimidines, three and four base-pair steps away from the cleavage site. Y76 is involved in an unusual stacking type interaction with a deoxyribose ring (T313), a major reason for distortions in the DNA. It is the intrinsically flexible substrate which has to adopt to the rigid framework of the enzyme structure.

The octamer duplex d(GGTATACC) (Octa V) is remarkably resistant to DNase I and, in contrast to Octa I, is not cleaved in the cocrystals. While the packing is different in the two crystal structures (continuous base stacking throughout the unit cell in the case of Octa V as opposed to discrete 14-mer units in Octa I), the general architecture and the overall DNA conformation is the same in both complexes (Fig. 2). The protein parts can be superimposed, with a root mean square (RMS) deviation of 0.7 Å for all atoms. The majority of the protein-DNA contacts are conserved in both structures; however, not surprisingly, there are additional contacts to the scissile phosphodiester bond which are missing in the Octa I complex. Tight interactions exist with N170, H134, H252, and D168 (Fig. 2). The disposition of residues around the phosphate suggests that H134 functions as a general acid protonating the leaving 03′, while H252 acts as a general base catalyst abstracting a proton from the attacking water molecule. Site-directed mutagenesis experiments show that both histidines are essential for activity (*19*, Connolly, personal communication). R9 was expected to form a salt bridge to the scissile phosphate, but is partially disordered as in the free enzyme and the Octa I complex and may play a role in stabilizing the transition state.

4. DNase I-induced DNA Conformation

X-ray structure analyses of short DNA fragments and their complexes with proteins clearly show the intrinsic flexibility of the DNA molecule (*25*, *26*). Not only the binding of a protein, but also the crystallization conditions and packing effects can influence and in some cases significantly distort the DNA (*27*, *28*). It is therefore dangerous to draw general conclusions from a particular DNA conformation present in a particular crystal structure. In the case of DNase I we are in the fortunate situation to be able to compare the conformation of two pieces of DNA with different sequence, present in a different packing environment, crystallized under different conditions (polyethylenglycol versus sodium citrate, pH 5.5 *versus* pH 4.4), and bound to an enzyme which does not change its conformation. Despite these differences, the overall conformation is the same in both complexes (Fig. 3). The common, local as well as global distortions of the DNA geometry must therefore be induced by DNase I. The DNA is bent away from the bound protein towards the major groove giving rise to a widened minor groove (Fig. 3). The minor groove width is approximately 15 Å in both complexes (as measured by the distance between nearest phosphorous atoms in opposite strands) as compared to about 12 Å in canonical B-DNA.

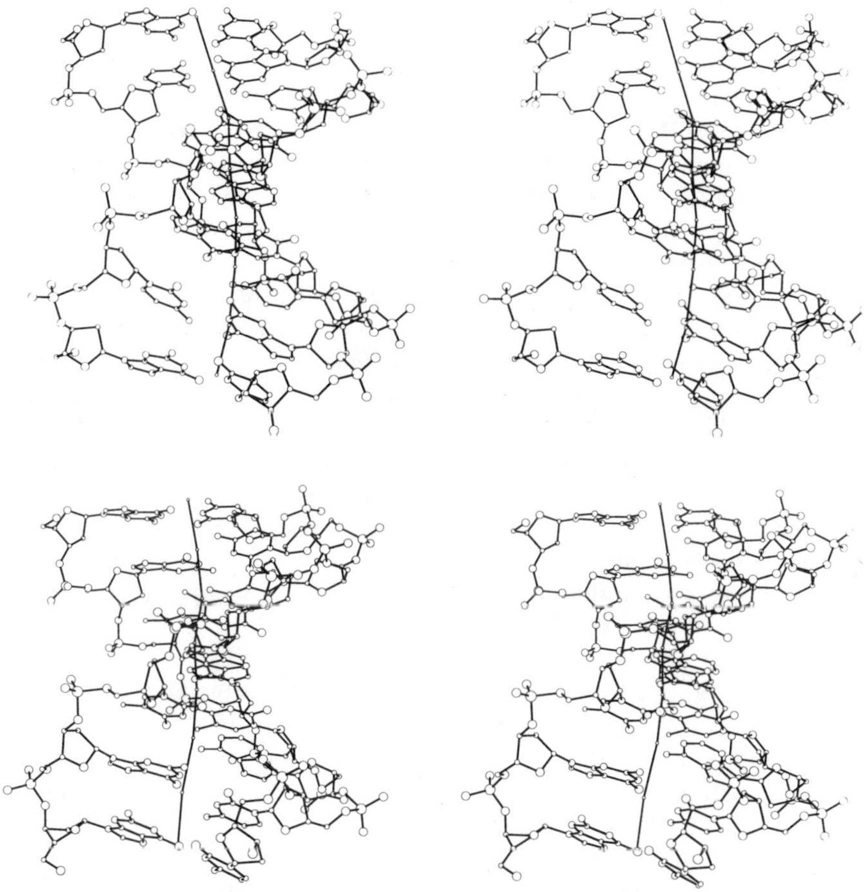

Fig. 3. DNase I-induced DNA conformations of Octa I (top) and Octa V (bottom). The minor groove in both duplexes (oriented to the right) is widened to about 15 Å and the helix is bent towards the major groove.

The most striking local distortion of the DNA is caused by the stacking interaction of Y76 with the deoxyribose ring of T313 (Fig. 2). In both complexes the torsional angles around the C4′-C5′ and C5′-05′ bonds of the thymidine are flipped from the usual *gauche/trans* to a *trans/gauche* conformation. A similar *trans/gauche* conformation next to the scissile phosphate (P315) in the Octa V complex is presumably induced by the multiple tight protein contacts at this position. The extensive base overlap at the step 312/313 is presumably a consequence of this backbone conformation.

In both complexes the helix parameters like, *e.g.*, tilt, rise, and twist as well as the apparent variability of sugar pucker and flexibility of backbone conformation are indicative of a B-type conformation. In Octa V, however, the lateral displacement values of the base-pairs from the helix axis are found approximately halfway between the values seen in canonical A- and B-form DNA. This displacement of the base-pairs leads to a shallower minor groove, which in turn will reduce the efficiency of binding to DNase I. It appears, therefore, that the remarkable resistance of the Octa V duplex compared to Octa I and other octamers is due to a narrower as well as shallower minor groove.

Other differences in local helix parameters clearly reflect the different base sequences of Octa I and Octa V. Of particular interest is the alternating stacking pattern with very little base overlap coupled to a high twist angle at TA steps and extensive base overlap and low twist at the AT steps of the d(GGTATACC)$_2$ duplex. Octa V shares these features with the crystalline d(CGCATATATGCG)$_2$ docecamer (*29*) and with the alternating B structure proposed for poly d(AT) (*30*).

5. *What Determines the Cleavage Rates of DNase I?*

The preceding description of two DNase I-DNA complex structures showed that binding by DNase I induces local as well as global distortions in the DNA substrate. The combination of tight interactions of a tyrosine (Y76) and an arginine (R41) in the minor groove with multiple contacts formed by both phosphate backbones leads to a widening of the minor groove and a bending towards the major groove of DNA. Hence the X-ray results seem to confirm earlier assumptions that the minor groove width is a major factor determining the cutting rates of DNase I (*3*, *18*). The narrow minor groove observed in X-ray structures of oligonucleotides with runs of A's or T's could therefore explain the very low DNase I cleavage rates in these regions (*31*, *32*). Figure 2 illustrates that the accommodation of side chains Y76 and R41 not only requires a certain width, but also a certain depth of the minor groove. An A-type helical conformation with a very shallow minor groove therefore hinders binding of DNase I. This explains why a DNA-RNA hybrid, known to exist in an A-conformation in solution is an extremely poor substrate for DNase I (Tinoco, personal communication). Consistent with this observation is the relative resistance of the Octa V duplex which has a significantly shallower minor groove than normal B-type oligonucleotides (see above).

The observed bending of the DNA towards the major groove indicates that not only does the actual size of the minor groove determine the susceptibility of a given DNA stretch, but also its bendability or stiffness. Using an elastic strain model, Hogan and Austin (*12*) calculated the variation of the DNA flexibility in an 80-bp-long DNA piece from a *Xenopus* control region and compared it to the observed cutting rates (*33*). They came to the conclusion that in the absence of significant local variations of DNA secondary structure, *i.e.*, in mixed-sequence DNA, DNase I cleavage is dominated by DNA flexibility variation. While we believe that more than one helical parameter is involved and therefore question the claimed quantitative agreement, the conclusion that flexibility is important for DNase I cutting rates is consistent with our X-ray results. It is conceivable that the lower cutting rates in G,C-rich regions is related to their lower flexibility compared to mixed-sequence DNA. Conformational stiffness may also add to the extraordinary insensitivity of poly A stretches.

The effect of bending on DNase I sensitivity is clearly visible in looped or circularly closed DNA. Compared to linear DNA of identical sequence, the susceptibility is enhanced in regions where the minor groove faces outwards and is therefore widened, but is decreased where it faces inside and is therefore compressed (*34*, *35*). The observed bending in the DNase I-DNA complexes also provides a possible explanation for the so-called hypersensitive sites (*36*). Sequences bent either intrinsically or through the action of a bound protein or drug molecule (*37*, *38*) towards the major groove would be particularly susceptible to DNase I degradation.

The parameters discussed so far, *i.e.*, groove width and stiffness, are slowly varying global helix parameters affecting several base pairs. These parameters cannot explain the often drastically different cutting rates at neighboring phosphodiester bonds. Other local parameters must be involved. The hypothesis that the cutting rate is proportional to the helical twist angle at the cleavage site (*11*) is not supported by our X-ray results. In fact, the scissile phosphodiester bond in Octa V as well as the purine-pyrimidine step preferentially cut in alternating purine-pyrimidine sequences is associated with a low twist angle. The Octa V complex suggests that the stacking and twisting pattern found in these alternating sequences is primarily responsible for the pronounced preference to cut 5′ to a pyrimidine. The tight packing in the minor groove suggests another possible local sequence effect: a G-C or C-G base pair instead of an A-T or T-A three base pairs away from the cleavage site would lead to unfavorably close contacts between the $2\text{-}NH_2$-groups of guanosine and R41 (Fig. 2). The lower than average cleavage rates of G,C-rich sequences may in part be due to these steric effects. Finally, as was already pointed out by Drew and Travers (*2*), it is likely that the orientation of the scissile phosphate group has an impact on the local cleavage rates.

II. P1 NUCLEASE

In contrast to DNase I, P1 nuclease from *Penicillium citrinum* shows strong preference for single-stranded substrates and cleaves DNA as well as RNA in an essentially unspecific reaction (*39*). P1 shares these properties with a number of zinc dependent nucleases, which are also able to recognize a range of non A/non B conformations in double-stranded nucleic acids (*40*). The enzyme is a phosphodiesterase cleaving the P-03′ bond, but at the same time acts as a 3′-nucleotidase. Both activities are strictly zinc-dependent and complete removal of the three zinc ions by EDTA treatment causes a loss of activity which is only partially restorable (*41*). This suggests an important function for the zinc ions not only in catalysis but also in stabilizing the folding of the enzyme. The 270 residue protein has four asparagine-linked carbohydrate units accounting for about 20% of the molecular weight (*42*). P1 is highly homologous to S1 nuclease from *Aspergillus oryzae*, showing approximately 50% sequence identity (*43*). The close relationship of the two enzymes is also evident from their enzymatic properties. Both enzymes require three zinc ions for activity, catalyze hydrolysis with inversion of configuration at the phosphorous atom and show the same stereo-selectivity for thiophosphorylated substrates (*44*).

The recently determined X-ray structure of P1 nuclease (*6*) provides the first atomic model of a zinc dependent, single strand specific nuclease. Based on the three-dimensional structure and results from crystal soaking experiments, I will discuss the interaction of P1 with single and double stranded substrates and the mechanism of their cleavage.

1. Three-dimensional Structure of P1 Nuclease

The three-dimensional fold of P1 is characterized by a large number of α-helices representing 56% of the structure (Fig. 4). There is only one short piece of β-structure present in the molecule. Two disulfide bonds and the three zinc ions bound at the bottom of the active site cleft further stabilize the structure. Two pairs of helices oriented along the long axis of the molecule form an approximately anti-parallel four helical bundle. While for three of the carbohydrate moieties density is visible only for the first sugar residue—a common feature for glycoproteins—there is significant density present for at least two N-acetylglucosamine and two mannose residues at the fourth glycosylation site at N92. At this position the carbohydrate packs against the protein shielding a tryptophane from the solvent.

An unexpected result of the X-ray analysis was the close structural similarity to phospholipase C (PLC) from *Bacillus cereus* (*45*). The two enzymes have essentially the same folding despite very limited sequence homology which is restricted to four short segments containing the zinc

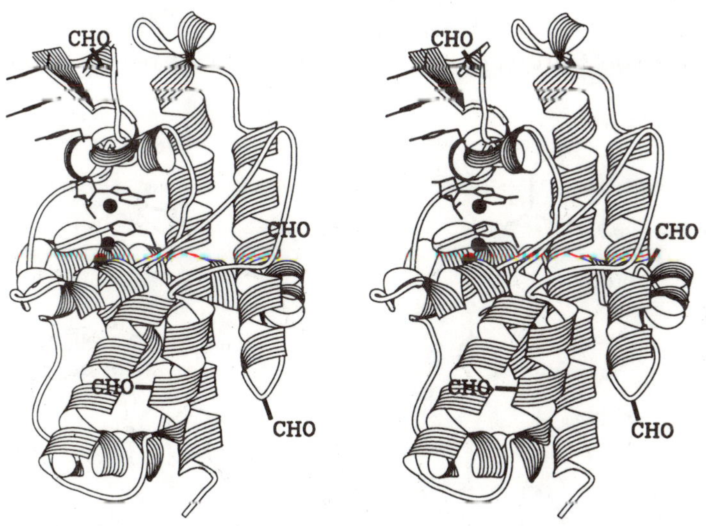

Fig. 4. Stereo ribbon plot of P1 nuclease. The three zinc ions bound in the active site cleft are indicated by black circles. A 5′-AMP is shown bound at the active site in stacking contact with F61. At a second site an adenine ring is stacked between two tyrosines. The four asparagine-linked carbohydrates are represented by the "CHO"-labels.

ligands. One hundred and forty-five equivalent C_α-atoms can be superimposed with a RMS deviation of 1.8 Å. Moreover, the arrangement and the coordination of the zinc ions is essentially the same in the two enzymes, including the presence of three water ligands. The similarities in folding, zinc coordination and function (both enzymes are phosphodiesterases) strongly suggest an evolutionary relationship.

2. *Coordination and Function of the Zinc Ions*

The zinc cluster in P1 is located at the bottom of the substrate binding cleft and consists of a relatively inaccessible dinuclear site with two zinc ions separated by 3.2 Å and a more exposed single site, roughly 5 Å removed from the other two (Fig. 5). Both zinc ions of the dinuclear pair are coordinated to four protein ligands and a bridging water molecule, while the more exposed zinc ion (Zn2) is liganded to only three amino acid ligands (H126, H149, D153) and two water molecules. The coordination of Zn2 resembles that observed in all so far structurally characterized catalytically active zinc sites (*46*), suggesting that it is directly involved in catalysis. The dinuclear zinc pair probably has an important function in stabilizing the folding of P1. It tightly links together three regions which are far apart in the amino acid sequence.

3. *Binding and Cleavage of Substrates*

A characteristic feature of the enzymatic activity of P1 is the strong

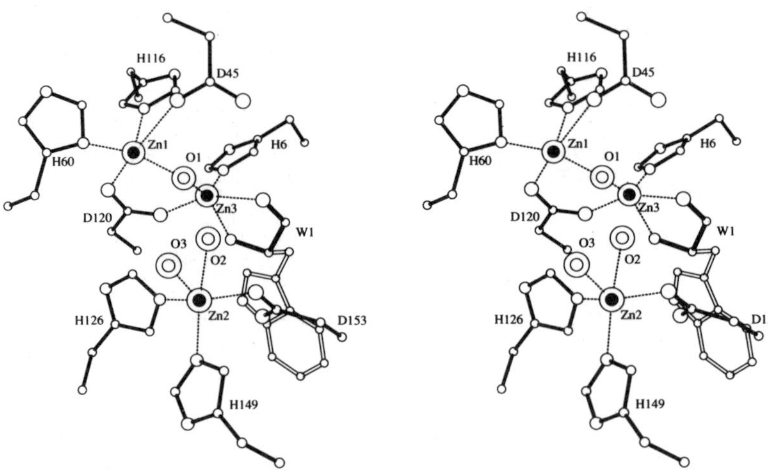

Fig. 5. Coordination of the zinc ions in P1 nuclease. 01, 02, 03 represent the water ligands.

preference for single-stranded nucleic acid conformations. The architecture of the molecule readily provides a clue for the origin of this preference: the catalytic zinc ion is located at the bottom of a cleft inaccessible to any double-stranded DNA or RNA. Docking of a canonical B-DNA into the substrate binding cleft reveals that the closest any phosphate group can approach to the catalytic zinc (Zn2) is about 7 Å. This suggests that the DNA has to be unwound before cleavage can occur.

The binding of substrates to P1 was studied by soaking tetragonal crystals with the R-diastereomer of the thiophosphorylated dinucleotide dAp(S)dA which is not cleaved by the enzyme (*44*). In a difference Fourier two well defined nucleotide binding sites were located, one at the active site close to the catalytic zinc and a second one approximately 20 Å away at the periphery of the molecule (Fig. 4). At both sites only a mononucleotide is visible in the difference density with the adenine ring bound in a hydrophobic pocket formed by F61 and V132 at the first and two tyrosine rings (Y144 and Y155) at the second site. At both sites hydrogen bonds of either the N6 or N1 positions of the adenine are possible with carboxylate groups (D63 and D146, respectively) at the bottom of the binding pockets. It is possible that the two exposed tyrosine rings (Y144 and Y155) form a second attachment site for a longer substrate. However, so far soaking experiments with a thiophosphorylated hexanucleotide only confirmed the two mononucleotide binding sites.

A comparison between P1 and other single stranded DNA binding proteins shows interesting common features. For example, the complex of RNase T1 and 2′-GMP like P1 shows a binding site in between two tyrosine side chains (*47*). Similar to P1, an exposed phenylalanine and a

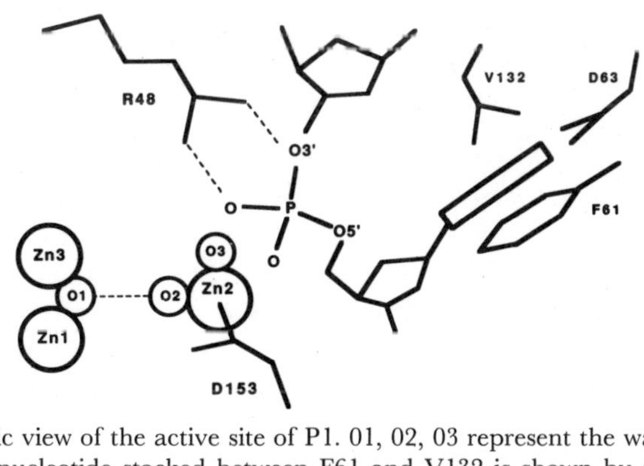

Fig. 6. Schematic view of the active site of P1. 01, 02, 03 represent the water ligands. The base of a bound nucleotide stacked between F61 and V132 is shown by a rectangle.

leucine form a hydrophobic binding pocket for a base at the 3′,5′-exonuclease site of DNA polymerase I (*48*, *49*).

The disposition of protein side chains and water molecules around the bound 5′-nucleotide indicates that the enzymatic reaction proceeds via a nucleophilic attack of a zinc activated water molecule (Fig. 6). The carboxylate group of D153, although already functioning as a zinc ligand, may assist in this activation. An assisting function of protein side chains in either orienting or activating the attacking water molecule has been proposed for a number of zinc dependent enzymes (*50*). R48, the only other side chain within hydrogen bonding distance of the phosphate, may serve to stabilize the pentacoordinate transition state.

SUMMARY

Bovine pancreatic DNase I shows a strong preference for double-stranded substrates and cleaves DNA with markedly varying cutting rates depending on the base sequence. The crystal structures of two complexes with octanucleotide duplexes show tight binding in the minor groove and to the sugar-phosphate backbones of both strands of the DNA. The combination of these contacts leads to distortions in the DNA, in particular to a widening of the minor groove and a bending towards the major groove suggesting that groove width and bendability are major parameters determining the cutting rates. While the general architecture, the type of interactions and the overall geometry of the DNA is highly similar in the two complexes, there are significant differences in local DNA structure which have to be attributed to the different sequences. The environment of the scissile phosphate in one of the complexes suggests that two His residues are involved in catalysis.

The single-strand specific P1 nuclease from *P. citrinum* cleaves RNA and DNA in an essentially unspecific reaction. The 3D-structure has been determined at 2.8 Å from a tetragonal and more recently at 2.2 Å from an orthorhombic crystal form. The predominantly α-helical protein has a fold very similar to that of phospholipase C despite very limited sequence homology. A cluster of 3 Zn ions is found at the bottom of the active site cleft, inaccessible to double-stranded DNA or RNA. Two nucleotide binding sites were identified by soaking the crystals with thiophosphorylated substrate analogues. The mode of binding suggests a catalytic mechanism involving nucleophilic attack by a Zn-activated water molecule.

Acknowledgments
I would like to thank my colleagues Armin Lahm, Christian Oefner, Anne Volbeda, and Simon Weston for their contributions to the work on P1 nuclease and DNase I. I gratefully acknowledge the expert help of Christina Kjaer in preparing the manuscript. The studies on DNase I were supported by grants from the Deutsche Forschungsgemeinschaft and the European Community.

REFERENCES

1. T.A. Steitz: *Q. Rev. Biophys.* **23**, 205–280 (1990).
2. H.R. Drew and A.A. Travers: *Cell* **37**, 491–502 (1984).
3. D. Suck, A. Lahm, and C. Oefner: *Nature* **322**, 465–468 (1988).
4. A. Lahm and D. Suck: *J. Mol. Biol.* **222**, 645–667 (1991).
5. A. Lahm, S.A. Weston, and D. Suck: *Nucl. Acids Mol. Biol.* **5**, 171–186 (1991).
6. A. Volbeda, A. Lahm, F. Sakiyama, and D. Suck: *EMBO J.* **10**, 1607–1618 (1991).
7. M. Laskowski: "The Enzymes" (Ed. by P.D. Boyer), pp. 289–311, Academic Press, New York/London (1971).
8. S. Moore: "The Enzymes" (Ed. by P.D. Boyer), pp. 281–296, Academic Press, New York/London (1981).
9. J.D. Galas and A. Schmitz: *Nucl. Acids Res.* **5**, 3157–3170 (1978).
10. A. Klug and L.C. Lutter: *Nucl. Acids Res.* **9**, 4267–4283 (1981).
11. G.P. Lomonossoff, P.J.G. Butler, and A. Klug: *J. Mol. Biol.* **149**, 745–760 (1981).
12. M.E. Hogan and R.H. Austin: *Nature* **329**, 263–266 (1987).
13. I. Brukner, V. Jurukowski, and A. Savic: *Nucl. Acids Res.* **18**, 891–894 (1990).
14. P.A. Price, T.Y. Liu, W.H. Stein, and S. Moore: *J. Biol. Chem.* **244**, 917–923 (1969).
15. S. Mehdi and J.A. Gerlt: *Biochemistry* **23**, 4844–4852 (1984).
16. D. Suck, C. Oefner, and W. Kabsch: *EMBO J.* **3**, 2423–2430 (1984).
17. C. Oefner and D. Suck: *J. Mol. Biol.* **192**, 605–632 (1986).
18. D. Suck and C. Oefner: *Nature* **321**, 620–625 (1986).
19. A.F. Worrall and B.A. Connolly: *J. Biol. Chem.* **265**, 21889–21895 (1990).
20. P.A. Price, W.H. Stein, and S. Moore: *J. Biol. Chem.* **244**, 929–932 (1969).
21. P.A. Price: *J. Biol. Chem.* **250**, 1981–1986 (1975).
22. S.R. Jordan, T.V. Whitcombe, J.M. Berg, and C.O. Pabo: *Science* **230**, 1383–1385 (1985).
23. S.C. Schultz, G.C. Shields, and T.A. Steitz: *J. Mol. Biol.* **213**, 159–166 (1990).
24. S.A. Weston, A. Lahm, and D. Suck: *J. Mol. Biol.* **226**, in press.
25. R.E. Dickerson: *Sci. Am.* **249**, 86–102 (1983).
26. Z. Shakked and D. Rabinovich: *Prog. Biophys. Mol. Biol.* **47**, 159–195 (1986).
27. J.A. McClarin, C.A. Frederick, B.C. Wang, P. Greene, H.W. Boyer, J. Grabble, and J.M. Rosenberg: *Science* **234**, 1526–1541 (1986).
28. A.D. DiGabriele, M.R. Sanderson, and T.A. Steitz: *Proc. Natl. Acad. Sci. U.S.A.* **86**, 1816–1820 (1989).
29. C. Yoon, G.G. Privé, D.S. Goodsell, and R.E. Dickerson: *Proc. Natl. Acad. Sci. U.S.A.* **85**,

6332–6336 (1988).
30. A. Klug, A. Jack, M.A. Viswamitra, O. Kennard, Z. Shakked, and T.A. Steitz: *J. Mol. Biol.* **131**, 669–680 (1979).
31. A.V. Fratini, M.L. Kopka, H.R. Drew, and R.E. Dickerson: *J. Biol. Chem.* **257**, 14686–14707 (1982).
32. H.C.M. Nelson, J.T. Finch, B.F. Luisi, and A. Klug: *Nature* **330**, 221–226 (1987).
33. M.E. Hogan, M.W. Roberson, and R.H. Austin: *Proc. Natl. Acad. Sci. U.S.A.* **86**, 9273–9277 (1989).
34. H.R. Drew and A.A. Travers: *J. Mol. Biol.* **186**, 773–790 (1985).
35. A. Hochschild and M. Ptashne: *Cell* **44**, 681–687 (1986).
36. S.C.R. Elgin: *Cell* **27**, 413–415 (1981).
37. Y. Huang, R.P. Rehfuss, S.R. La Plante, E. Boudreau, P.N. Borer, and M.J. Lane: *Nucl. Acids Res.* **16**, 11125–11139 (1988).
38. R. Mendoza, J. Markovits, J.P. Jaffrezou, G. Muzard, and J.B. Le Pecq: *Biochemistry* **29**, 5035–5043 (1990).
39. K. Shishido and T. Ando: In "Nucleases" (Ed. by S.M. Linn and R.J. Roberts), pp. 155–185, Cold Spring Harbor Laboratory Press, Cold Spring Harbor, New York (1982).
40. D.E. Pulleyblank, M. Glover, Ch. Farah, and D.B. Haniford: In "Unusual DNA Structures" (Ed. by R.D. Wells and S.C. Harvey), pp. 23–44, Springer-Verlag, Heidelberg (1988).
41. M. Fujimoto, A. Kuninaka, and H. Yoshino: *Agric. Biol. Chem.* **38**, 785–790 (1974).
42. K. Maekawa, S. Tsunasawa, G. Dibó, and F. Sakiyama: *Eur. J. Biochem.* **200**, 651–661 (1991).
43. A. Iwamatsu, H. Aoyama, G. Dibó, S. Tsunasawa, and F. Sakiyama: *J. Biochem.* **110**, 151–158 (1991).
44. B.V.L. Potter, B.A. Connolly, B.A., and F. Eckstein: *Biochemistry* **22**, 1369–1377 (1983).
45. E. Hough, L.K. Hansen, B. Birknes, K. Jynge, S. Hansen, A. Hordvik, C. Little, E. Dodson, and Z. Derewenda: *Nature* **338**, 357–360 (1989).
46. B.L. Vallee and D.S. Auld: *Proc. Natl. Acad. Sci. U.S.A.* **87**, 220–224 (1990).
47. R. Arni, U. Heinemann, R. Tokuoka, and W. Saenger: *J. Biol. Chem.* **263**, 15358–15368 (1988).
48. L.S. Beese and T.A. Steitz: *EMBO J.* **10**, 25–33 (1991).
49. P.S. Freemont, J.M. Friedman, L.S. Beese, M.R. Sanderson, and T.A. Steitz: *Proc. Natl. Acad. Sci. U.S.A.* **85**, 8924–8928 (1988).
50. H. Kim and W.N. Lipscomb: *Biochemistry* **29**, 5546–5555 (1990).

Molecular Structure and Life (Ed. by Y. Kyogoku and Y. Nishimura), pp. 157–170, Japan Sci. Soc. Press, Tokyo/CRC Press, Boca Raton (1992)

DNA-binding Protein Motifs: Several Repeats and a Palindrome

Masashi SUZUKI

*MRC Laboratory of Molecular Biology**

The laws of physics are simple and beautiful. When molecular biology was born 40 years ago, scientists expected the same simplicity in biology. Unfortunately, textbooks of molecular biology are becoming more and more complicated. Even physics is beautiful only if extremely simple subjects are chosen; thus in Newtonian mechanics we can handle only a few particles perfectly, and in Boltzmann's statistical mechanics we can describe the behaviour of billions of identical particles only as a mass. The reality lies between the two extremes, where physics is less powerful and less beautiful. Biologists cannot escape from this complicated reality.

The laws of the whole of biology cannot be written down in three sentences. But this does not mean that rules do not exist. The rules in biology are much more sophisticated and by their nature resemble grammar: the rules of a language. How can we "decipher" such sophisticated rules of DNA-recognition by proteins ? Some biophysicists believe that the rules of protein folding are not as complicated as once believed. Already several structural frameworks shared by many proteins have been discovered. These "motifs" are the units of structure and function

* Cambridge CB2 2QH, U.K.

of proteins. Some DNA-binding proteins contain odd sequences which occur repeatedly, or even palindromically. The "oddness" in these sequences reflects restrictions in designing DNA-binding units to recognise features of binding sites on DNA. Thus these sequences provide clues for identifying the regions of DNA-binding proteins involved in DNA-recognition and understanding their structures and their modes of DNA-binding. Here, I discuss some of these sequences.

I. SPKK MOTIF

The SPKK motif was originally identified as a repeating unit of the N-terminal sequences of sea urchin histones H1 and H2B (*1*). Variants of this motif have been found in many DNA-binding proteins (*2–4*). Histone H1 is composed of three domains, the two hydrophilic termini and the central globular domain (reviewed in (*5*)). Sea urchin spermatogenous H1 has a strange repeat in the N-terminal sequence (also H2B). After a few groups tried to figure out the repeating unit and its conformation (*6*, *7*), I identified the repeating unit as Ser-Pro-(Lys/Arg)-(Lys/Arg) rather than another combination such as KKSP (*1*).

I was able to identify the unit based entirely on personal experience, because I once studied a methylation motif, APKK, where the N-terminal Ala is a trimethylation site (*8*, *9*). The trimethyl-APKK sequence occurs in the N-terminal ends of myosin light chains and histone H2B of tetrahymena and insects. The recognition of this sequence by a monoclonal antibody (SI304) is independent of differences in the sequences following APKK (*8*, *9*). Thus, this four amino acid sequence is likely to fold into an independent structure. The methylation of this sequence in *Drosophila* H2B occurs as a heat shock response (*10*). The X-P-K-K sequence, where X is a modification site, was therefore familiar to me when I came across the sea urchin histone sequences. I recognised the SPKK sequence as a structural/modification motif similar to the APKK motif, having Ser as a phosphorylation site.

A structure which can be formed by four amino acid residues is a β-turn. I have proposed that the structure of the SPKK motif is a β-turn I with an additional hydrogen bond between the side chain oxygen atom and the main chain amide of the third residue ($\beta\sigma$-turn) (*1*). This idea, the β-turn formation by the SPKK motif, was supported by statistical analysis on protein crystal structures (*2*, *11*), and tested by circular dichroism (CD) ((*1*), Suzuki, manuscript in preparation), NMR (Suzuki and Neuhaus, submitted) and infrared (IR) (*12*) studies using synthetic peptides.

About half of the (Ser/Thr)-Pro-X-X fragments in protein crystal structures fold into a β-turn or one of a few closely related structures. These structures are characterised by differences in the way the three hydrogen bonds are combined: a β-type between CO of Ser/Thr and NH of the fourth residue, a σ-type between OH of Ser/Thr and NH of the third residue, and a τ-type between OH of Thr and NH of the fourth residue. The most favored combination is $\beta\sigma$ for the SPXX (Fig. 1a) and $\beta\tau$ by TPXX (Fig. 1b) (*11*).

The Thr-Pro-Arg-Lys peptide which constitutes one unit of the SPKK motif was studied by NMR measurements (Suzuki and Neuhaus, submitted). In dimethyl sulfoxide the TPRK peptide adopts one of two primary, slowly-exchanging conformations, a major one (M) in which the proline is *trans*, and a minor one (m) in which the proline is *cis*. The M conformation is itself a mixture of extended and structured states in rapid flux. The structured state is likely to be a β-turn in view of the low

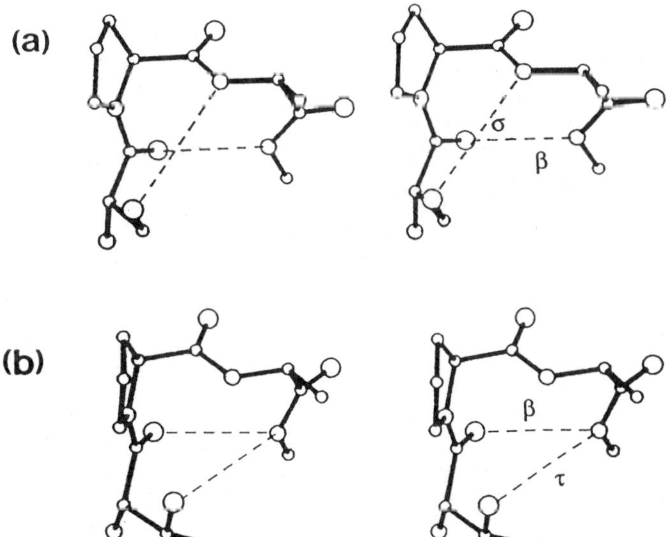

Fig. 1. Stereoviews of (Ser/Thr)-Pro-X-X fragments in protein crystals. Standard structures of a $\beta\sigma$-turn of a SPXX fragment (a) and a $\beta\tau$-turn of a TPXX fragment (b) are shown (*11*). Three types of hydrogen bonds, β, σ, and τ are indicated by broken lines.

temperature dependence of the Lys amide proton resonances, and two NOE (nuclear Overhauser effect) connectivities, one between the backbone NH protons of Lys and Arg, the other between Pro H_α and the Lys backbone NH (Fig. 2a) (Suzuki and Neuhaus, submitted). Figure 2b

(a)

Major conformation NOE cross peaks	Minor conformation NOE cross peaks	Conformational Exchange
1 P H_α - R NH(S)*	11 P H_α - R NH*	101 T H_α
2 R H_α - K NH(S)*	12 R H_α - K NH*	102 P H_α*
3 T H_α - P H_δ(S)*	13 T H_α - P H_α	103 R H_α
4 R NH - K NH(S)*	14 R NH - K NH	104 R NH
5 P $H_{\beta A}$ - R NH(S)*	15 P $H_{\beta A}$ - R NH	105 K NH
6 P $H_{\beta B}$ - R NH(S)*		106 K H_ε*
7 P $H_{\beta A}$ - K NH(w)*	17 P $H_{\beta A}$ - K NH	
8 P H_α - K NH(w)		
9 T $C_\gamma H_3$ - P H_δ(w)*	19 T $C_\gamma H_3$ - P H_α	

(b)

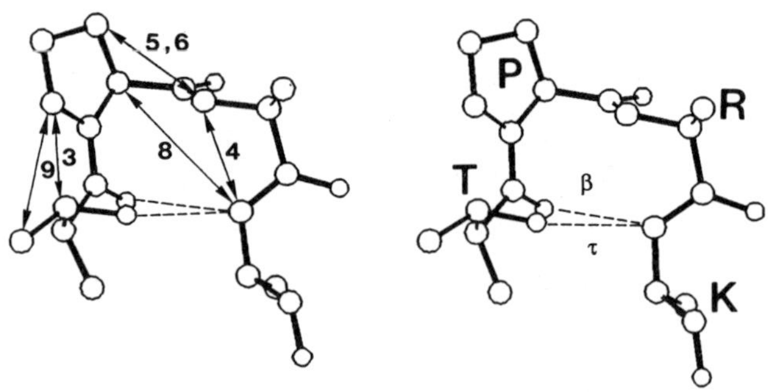

Fig. 2. NMR structure of the TPRK peptide. (a) Observed NOE and conformational exchange cross peaks arising from species M and m. Only inter-residual NOE cross peaks are listed here. Signals confirmed by ROESY are marked *. Intensities are indicated as strong (S) or weak (w) (Suzuki and Neuhaus, submitted). (b) Stereoview of the proposed structure of the M conformation of TPRK, based on NMR analysis (Suzuki and Neuhaus, submitted). For clarity, the side chain beyond C_β is omitted for Arg and Lys. The proposed bifurcated hydrogen bond, which is the combination of τ-type (τ) and β-type (β), involving Lys NH, Thr CO, and Thr side-chain OH is indicated by broken lines. Arrows indicate selected NOE connectivities, numbered as in (a).

shows a model of the proposed M conformation of the TPRK peptide derived from the NMR analysis (Suzuki and Neuhaus, submitted).

While it is clear that the peptide is in a very different environment in DMSO solution to that which it would experience when bound to DNA, the NMR analysis suggests that the SPKK type sequence has some intrinsic tendency to fold into a β-turn conformation, and such a structure is therefore a good candidate for the conformation adopted by this peptide in its complex with DNA.

The DNA-binding mode of the SPKK motif has been studied in solution by fluorescence measurements and footprinting experiments (*1*, *13*). The SPKK motif binds to A/T-rich sequences in the minor groove of DNA as do DNA-binding drugs such as netropsin, Hoechst 33258, *etc.* A study using computer graphics has shown that repeats of β-turns can fit into the minor groove of DNA and can bind to A/T base pairs by hydrogen bonds using the *i*-th NH's (*1*).

This minor groove binding model seems to be essentially correct. The actual sequences of the SPKK motif repeats vary slightly from each other, and can be classified into three types:
(A) XX**SPQK**RAA[**SPRK**]$_{n=3-5}$ASA**SPRR**KAK,
(B) XX**SPSK**G[**SPRKG**]$_{n=2-3}$**SPRK**KA**SPKR**GGKGA,
(C) PSQK[**SPTKR**]$_{n=3}$**SPRK**GGRGA.
Type A is found in all echinoderm spermatogenous H1, while type B occurs in the major component of sea urchin H2B and feather star H2B, and type C in the minor component(s) of sea urchin H2B and in the dominant component of echinothulid sea urchin H2B (Suzuki *et al.*, unpublished results). The three sequences likely result from differences in the way the SPKK motifs are physically connected so that they can follow the minor groove of DNA.

II. SEGMENTED HELIX

Another interesting repeat sequence is found in the *Drosophila* K10 gene product (*14*). This sequence is composed of eight repeats of the octapeptide, SPNQQQHP, or its closely related variants (Fig. 3a) ((*4*), Suzuki and Aimoto, submitted). The sequence is extremely well designed to fold into a repeat of short helical blocks to wrap around a DNA strand. If a long α-helix were to bind to DNA, it would presumably have to be kinked and separated into short blocks (*15*) in order to follow the groove of DNA. The curvature would be too great to permit a continuous run of hydrogen bonds.

The PH2 peptide, a peptide which constitutes the two K10 standard repeats, binds to A/T-rich sequences (Suzuki and Aimoto, submitted). The DNA-binding constant of PH2 is estimated as 2.0×10^5 M^{-1}. One PH2-binding site seems to be eight or nine base pairs long and thus one unit covers about four base pairs.

The CD spectrum of PH2 in an aqueous solution is characterised by two negative peaks (Suzuki and Aimoto, submitted), one close to 200 nm, which reflects an unstructured state, and the other at 222 nm, which reflects the presence of a helical conformation (*16*). On adding increasing amounts of trifluoroethanol (TFE), which is known to stabilise peptide secondary structures, the mean residue ellipticity at 222 nm (θ_{222}) becomes larger, and θ_{200} smaller. This is what is expected for a coil-to-helix transition.

However, the CD spectrum of PH2 in TFE is not that of a typical α-helix. The isospectroscopic point of typical coil-to-helix transition is 204 nm (*16*), while that of PH2 is 212 nm. Moreover, a typical α-helix has

(a)

SPKKKSMK
GIKQQQHP
SPNQQQPP
SPNQQQHP
SPNQQQHP
SPNQQQHP
NSNQQQHP
SPNQQQGK

(b)

SPKKAKKP	*AAAATK*
SPKKAKKP	*AAAATK*
SPKKAKKP	*AAAATK*
SPKKAKKP	*VKAAKS*
SPKKAKKP	*VKAAKS*
SPKKAKKP	*KTVKAK*
SPKKAKKP	*KTVKAK*
SPKKAKKP	*KTVKAK*
SPKKAKKP	*KTVKAK*
SPKKAKKP	*AKKSPK*
SPKKAKKP	*ATPKKA*
SPKKAKKP	*ATPKKA*
SPKKARK–	
SPKK*V*KKP	*AAAAKK*
SPKK*V*KKP	*AKAALA*
SPKK*V*KKP	*AAAKK*–
SPKKAKK*A*	*AGKRKP*
SPKKAARP	*RKARSP*
SPKKAARP	*RKARSP*
SPKKAARP	*RKARSP*
SPKKA*VK*P	*KAAKPK*

SPKKA*TKAAKPKAA*
SPKKA*TKAAKPKAA*
SPKKA*TKAGRPKTT*
SPKKA*AKAGRPKKA*
SPKKA*AKAGRPKKA*
SPRKA*GKRRSPKKA*

Fig. 3. Sequences of the K10 octapeptide repeat (a) and some SPXX type fragment sequences in the C-termini of various histone H1's (b). Sequences are according to (*14*, *23*).

another negative peak at 208 nm (*16*), which is not clearly observed with PH2. The spectra of peptides containing one, two, and four repeats are similar and the thermodynamical characteristics of these peptides are almost the same. This contrasts with the case of a pure α-helical structure which becomes more stable as the number of amino acid residues included increases (*17*). Thus the structure of a repeating unit seems independent of its neighbours, and the coil-to-helix transition does not propagate all the way along the molecule.

Since it is unlikely that the His-Pro-Ser-Pro sequence, which contains two Pro residues, folds into an α-helix, this would be expected to form the connection between the helical blocks, while the Q cluster would form the core of the helix. A serine residue at the i-th position can cap a helix at its N-terminus by a hydrogen bond to the $i+2$-th NH (N-cap, (*18*)). The SPXX sequence is especially suitable for capping a helix, because this sequence has a high tendency to fold into a $\beta\sigma$-turn (*2*, *11*), in which the $i+2$-th and $i+3$-th NHs are capped by the σ type hydrogen bond, from OH of Ser to the $i+2$-th NH, and the β-turn type hydrogen bond, from CO of Ser to the $i+3$-th NH, respectively; also, the $i+1$-th Pro lacks NH. Consequently, if such a structure is formed at the N-terminus of a helix, the coil-to-helix transition cannot propagate towards the N-terminus. Among the 10 β/σ-turns adopted by the (S/T)PXX sequences in PDB (*11*), six examples are located at the N-terminus of an α-helix. We therefore have proposed that this structure, the $\beta\sigma$-turn incorporated into the N-terminus of a helix, forms the K10 repeat unit (Suzuki and Aimoto, submitted) (Fig. 4a).

The HPSP junction can cap the C-terminus of the helix as well (C-cap). If the SP sequence of HPSP folds into a $\beta\sigma$-turn, statistics suggest that the dihedral angles of His-Pro-Ser should become β-strand-like, because of restricted possibilities in the XPXP conformations (M. MacArthur and J.M. Thornton, personal communication). Their statistics suggest a strong conformational restriction in XPXP structure and that there are essentially four possible combinations of dihedral angles for the PXP structure. If the SP part of the HPSP sequence folds into a β-turn I, it forms part of the set, β-β-α, where β and α, respectively, mean that the dihedral angles resemble that of β-strand and that of α-helix. In almost all examples in their statistics in which this combination occurs, the dihedral angles of the residue preceding PXP are also β-strand-like. Thus the dihedral angles of HPSP are likely to be β-β-β-α. The β-strand region separating two helical blocks again appears to stop the propagation of coil-to-helix transition.

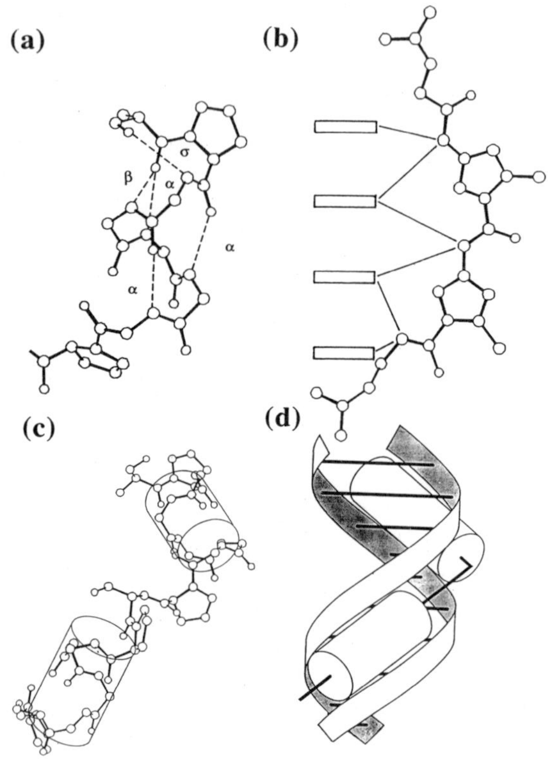

Fig. 4. Segmented helix structure. The models proposed for one unit, SPNQQQHP, (a) and two units, $[SPNQQQHP]_2$, (c), of the K10 repeat are shown (Suzuki and Aimoto, submitted). Broken lines show hydrogen bonds of the α-helix type (α), the β-turn type (β), and the σ-turn type (σ). Side chains of Asn, Gln, and His are omitted. (b) The structure of netropsin binding to DNA (*19*) is shown in the same scale as used for (a). DNA base pairs binding netropsin are schematically drawn as boxes. (d) A schematic drawing of DNA-binding by the two SPNQQQHP units is shown.

The proposed model of two units is shown in Fig. 4c. It is of the right form and has sufficient degree of freedom to follow the groove of DNA (Fig. 4d). The length of one unit, SPNQQQHP, is slightly shorter than that of netropsin which covers about five base pairs in the minor groove of DNA (Fig. 4b, (*19*)). Thus, the proposed one unit structure probably covers four base pairs, and this agrees with the results of footprinting experiments.

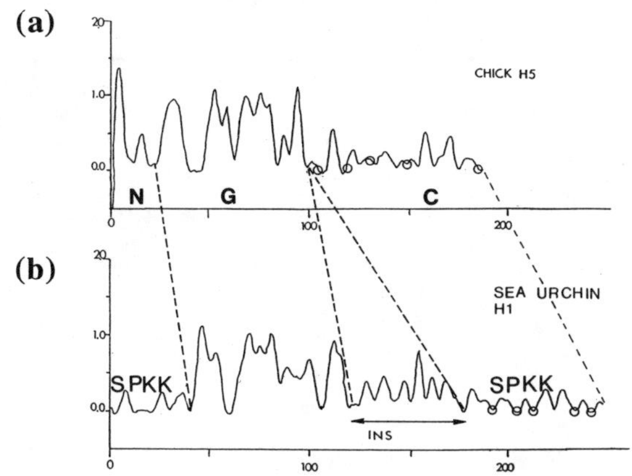

Fig. 5. Insertion region in sea urchin histone H1. Plots of hydrophobicity of chicken H5 (a) and sea urchin H1 (b) are compared. The histone sequences are according to (*23*). Histone H1/H5 is composed of three parts, N-terminal (N), central globular (G), and C-terminal (C). The C-terminal parts of the two histones contain the SPKK motifs (marked with circles). About 60 amino acid residues (residues 122–176) are inserted between the globular domain and the C-terminus in sea urchin H1. The N-terminus of sea urchin H1 is composed of the SPKK motifs.

III. PALINDROMIC α-HELIX

The segmented helix is a protein structural motif designed to follow the continuity of the same DNA double helix by breaking the continuity of a helix. In contrast, a continuous α-helix, if it is sufficiently long, might be expected to bridge two DNA double helices or more. Such a motif seems to exist, to my surprise, as a palindromic sequence (*20*). In sea urchin spermatogenous H1 a stretch of amino acid residues is inserted between the globular domain and the C-terminus (*20, 21*) (Fig. 5), and it has been shown that the insertion has a propensity to fold into an α-helix (*21*). The insertion in H1 of the species, *Parechinus angulosus*, is 55 amino acids long (residues 122–176). Surprisingly, the 37 amino acid sequence in the insertion (residues 140–176), is **palindromic**, that is, the sequence reads the same way in the C to N direction as in the N to C, provided minor differences in one case between Lys and Arg, and in one case between Ala and Leu are neglected (Suzuki, submitted) (Fig. 6a). If this 37 amino acid sequence is plotted onto an α-helical projection, one face of the α-helix is covered only by Ala, while all Arg are located on the opposite face.

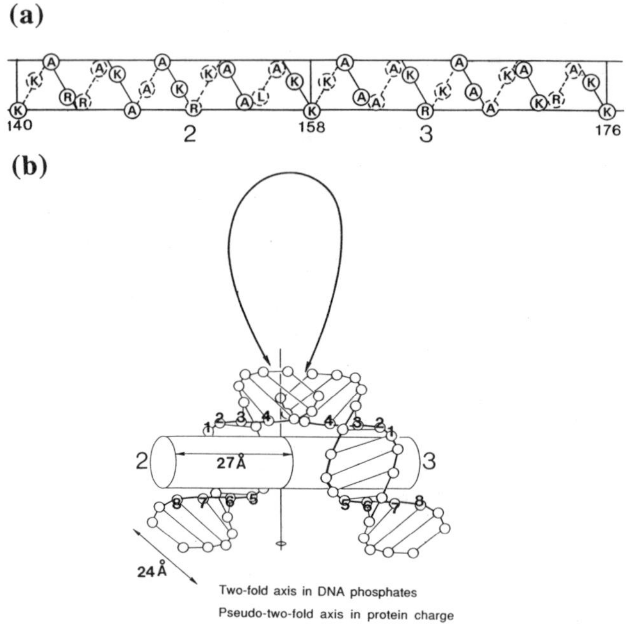

Fig. 6. Palindromic α-helix. The palindromic sequence in H1 (residues 142–176) is plotted onto an α-helical diagram (a). One surface of the α-helix is covered only with Ala, while all Arg are located on the opposite surface. Each half of the 37 amino acid sequence, (2) and (3), spans 27 Å, if folded into an α-helix (b). If such an α-helix is placed into the groove of DNA, of which the diameter is 24 Å, it would cover about half a double helical turn of DNA, which has 8–10 phosphate groups (b). The α-helix folded by the palindromic sequence has a pseudo two-fold axis in its structure (b). Two DNA strands binding to it in the same manner would be related by the same two-fold axis. Thus the palindromic helix may well stabilise one superhelical DNA turn (b).

Each half of the 37 amino acid sequence, which I call (2) and (3) respectively, spans 27 Å, if folded into an α-helix (Fig. 6b). If such an α-helix is placed in the groove of B-DNA, which has a diameter of 24 Å, it will cover about half a double helical turn of DNA. The number of basic residues (8–9) in each α-helix, is close to the number of phosphates in two phosphate helices of one DNA double helical strand accessible from the α-helix (8–10). This geometry and the charge distribution of the α-helix strongly suggest that each 19 residues constitutes an independent DNA-binding domain by folding into an α-helix.

Since the α-helix folded by the palindromic sequence has a pseudo two-fold axis in its structure, two DNA strands binding to it in the same manner would be related by the same two-fold axis (Fig. 6b). The length of the linker DNA connecting two neighbouring nucleosomes varies among species and that of sea urchin is about 80 bp long, a value close to the number of base pairs in one superhelical turn around the histone octamer (*22*). Thus it is possible that the palindromic α-helix, which is not found in H1 of other organisms, stabilises the formation of an additional superhelical turn formed by the 80 bp linker DNA using its palindromic nature (Fig. 6b).

IV. H1 C-TERMINUS

The segmented helix, which is the model proposed for the K10 repeat, might have further relevance. The SPKK motifs present in the C-terminus of histone H1 are separated by some amino acid residues (see the histone sequences listed in (*23*)). The C-terminus seems to be composed of 7–8 subdomains and some of them start with the Ser-Pro sequences (see the periodicity in the hydrophobicity of the region in Fig. 5b). In the H1 C-terminus sequences listed by von Holt *et al.* (*23*) 50 SPXX sequences are found and 28 out of the 50 are of the SPKK(A/V)X(K/R) type. Twenty-one sequences amongst them are SPKKAKKP or closely related sequences (Fig. 3b). The SPKKAKKP sequence starts with SP and ends with Pro at the 8-th position like the SPNQQQHP sequence. The synthetic peptide, KSPKKAKKP, shows a CD spectrum similar to that of PH2 (Suzuki, unpublished). If the SPKKAKKP is helical, the N-terminus is properly capped by the SPKK motif, but the C-termini might not be capped but twisted as is frequently seen in Pro-containing α-helices. Thus, although the H1 C-terminus is not composed of a very regular repeat of a sequence, it seems to fold into a structure suitable to wrap around DNA by being broken into segments of helix. Two groups have proposed different secondary structures for the H1 C-terminus, an α-helix (*24*) and a β-turn I (*12*), but I believe that the region folds into a combination of both.

SUMMARY

The **SPKK motif** is found in histones H1 and H2B. The structure ($\beta\sigma$-turn formation) and the DNA-binding mode (minor groove-binding) of the SPKK motif repeat are discussed.

The *Drosophila* K10 repeat is composed of eight octapeptides, **SPNQQQHP** or its variant sequences. On the basis of CD and footprinting

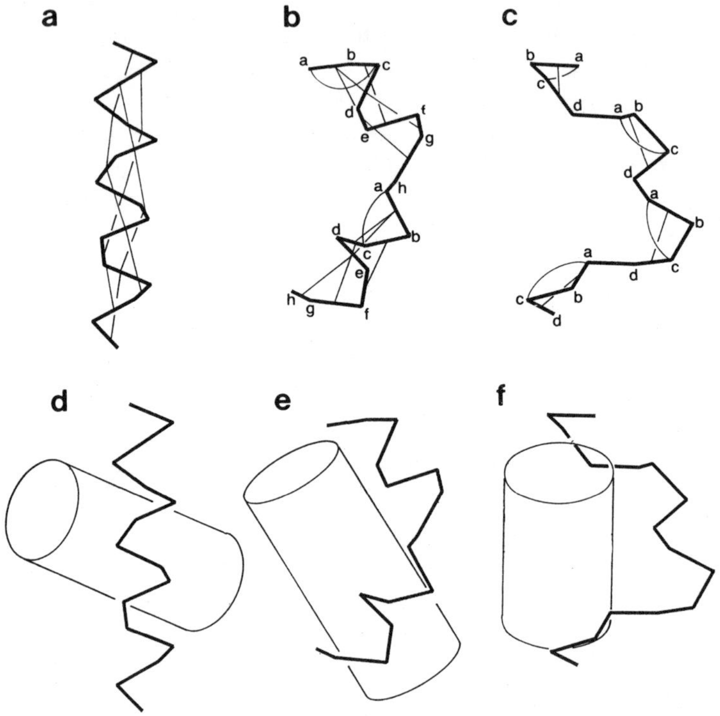

Fig. 7. Comparison of structures of DNA-binding motifs. The structures of a standard α-helix, which is used for many DNA-binding motifs including the palindromic α-helix, (a), (d), the segmented helix proposed for the K10 repeat, (b), (e), and repeats of β-turns (I) proposed for the SPKK repeat, (c), (f), are compared. Each of the three structures shown here is composed of 16 amino acid residues. Only the C_α connections are shown. In (a)–(c) hydrogen bonds are also indicated. In (d)–(f) DNA is schematically represented by cylinders. Each unit of the helical segments in (b) and each β-turn unit in (c) is indicated with characters, (a)–(h) and (a)–(d), respectively.

studies, a segmented helix model, which is a helix broken into short pieces to follow the groove of the same DNA double helix, is proposed. This model also seems to have relevance to the structure of the H1 C-terminus.

The palindromic 37 amino acid sequence, **KK**A**RR**A**K**AAA**KRK**AALA**K-K-K**AAAA**KRK**AAA**K**A**KR**A**KK**, is found in an α-helical region of histone H1. It is proposed that this motif bridges two DNA double helices in a symmetrical way and stabilises a superhelical configuration of the DNA by folding into a **palindromic α-helix**.

The structure of the segmented helix is a combination of two structures, helix and β-turn, and thus the three structures which appear to be used for DNA-recognition, the continuous α-helix, the segmented helix, and the repeated β-turns, are related to each other in their structures (Fig. 7a–c) and in their DNA-binding modes (Fig. 7d–f).

Acknowledgments
I thank Drs. C. Chothia, J. Finch, D. Rhodes, J.A. Grasby, and A. Klug for reading the manuscript, critical comments and continuous encouragement.

REFERENCES

1. M. Suzuki: *EMBO J.* **8**, 797–804 (1989a).
2. M. Suzuki: *J. Mol. Biol.* **207**, 61–84 (1989b).
3. M. Suzuki: *Nature* **344**, 562–565 (1990).
4. M. Suzuki: In "Nucleic Acids and Molecular Biology 5" (Ed. by F. Eckstein and D.M. J. Lilley), pp. 126–140, Springer-Verlag, Berlin-Heidelberg (1991).
5. E.M. Bradbury, N. Maclean, and H.R. Matthews: "DNA, Chromatin, and Chromosomes," Blackwell, Oxford (1981).
6. C. von Holt, P. de Groot, S. Schwager, and W.F. Brandt: In "Histone Genes" (Ed. by G.S. Stein, J.L. Stein, and W.F. Marzluff), pp. 65–105, Wiley-Interscience, New York (1984).
7. D. Poccia: In "Molecular Regulation of Nuclear Events in Mitosis and Meiosis" (Ed. by R.A. Schlegel and M.S. Halleck), pp. 149–178, Academic Press, Orlando, Florida (1987).
8. M. Tokunaga, M. Suzuki, K. Saeki, and T. Wakabayashi: *J. Mol. Biol.* **194**, 245–255 (1987).
9. M. Suzuki, K. Saeki, M. Tokunaga, and T. Wakabayashi: *Proc. Jpn. Acad.* **63 B**, 385–388 (1987).
10. R. Desrosiers and R. Tanguay: *J. Biol. Chem.* **263**, 4686–4692 (1988).
11. M. Suzuki and N. Yagi: *Proc. Roy. Soc. Lond. B* **246**, 231–235 (1991).
12. M. Erard, F. Lakhdra-Ghazal, and F. Amalric: *Eur. J. Biochem.* **191**, 19–26 (1990).
13. M.E.A. Churchill and M. Suzuki: *EMBO J.* **8**, 4189–4195 (1989).
14. E. Prost, F. Deryckere, C. Roos, M. Haenlin, V. Pantesco, and E. Mohier: *Genes Dev.* **2**, 891–900 (1988).
15. R.W. Warrant and S.-H. Kim: *Nature* **271**, 130–135 (1978).
16. N. Greenfield and G.D. Fasman: *Biochemistry* **8**, 4108–4116 (1969).
17. B.H. Zimm and J.K. Bragg: *J. Chem. Phys.* **31**, 526–535 (1959)
18. I.S. Richardson and D.C. Richardson: *Science* **240**, 1648–1652 (1988).
19. M.L. Kopka, P.E. Pjura, D.E. Goodsell, and R.E. Dickerson: *Proc. Natl. Acad. Sci. U.S.A.* **82**, 1376–1380 (1986).
20. M. Suzuki, *Proc. Natl. Acad. Sci. U.S.A.*, in press.
21. C.S. Hill, S.R. Martin, and J.O. Thomas: *EMBO J.* **8**, 2591–2599 (1989).

22. T.J. Richmond, B. Rushton, D. Rhodes, and A. Klug: *Nature* **311**, 532–537 (1984).
23. C. von Holt, W.F. Brandt, H.J. Greyling, G.G. Lindsey, J.D. Retief, J. de A. Rodriguess, S. Schwager, and B.T. Sewell: *Methods Enzymol.* **170**, 431–523 (1989).
24. D.J. Clark, C.S. Hill, S.R. Martin, and J.O. Thomas: *EMBO J.* **7**, 69–75 (1988).

MOLECULAR STRUCTURE IN GENE REGULATION

Molecular Structure and Life (Ed. by Y. Kyogoku and Y. Nishimura), pp. 173–182, Japan Sci. Soc. Press, Tokyo/CRC Press, Boca Raton (1992)

Selective Regulation of Transcription by Members of the C/EBP Family

Peter LAMB and Steven L. MCKNIGHT

*Howard Hughes Research Laboratories, Carnegie Institution of Washington, Department of Embryology**

Specific interactions between proteins and DNA have long been known to be central to the process of transcriptional control. Recent studies have shown that specific protein-protein interactions are often intimately involved in the process of protein-DNA interaction, as evidenced by the large number of transcriptional regulatory proteins that bind as dimers. In this paper, we present a hypothesis that proposes an additional role for specific protein-protein interactions, in this case between adjacent DNA-bound transcription factors. These interactions are postulated to increase the specificity with which transcriptional regulatory proteins bind to promoters and enhancers.

The bZIP class of transcription factors constitutes a large family of DNA binding proteins that employ a common architecture for interacting with specific sites on DNA (*1*). The bZIP domain is separable into two components—a region rich in basic amino acids that forms the DNA contact surface, and a leucine zipper that acts as a dimerization interface (*2*, *3*). The leucine zipper consists of an amphipathic α-helix that serves to bring together two protein monomers such that their basic regions are

* Baltimore, MD 21210, U.S.A.

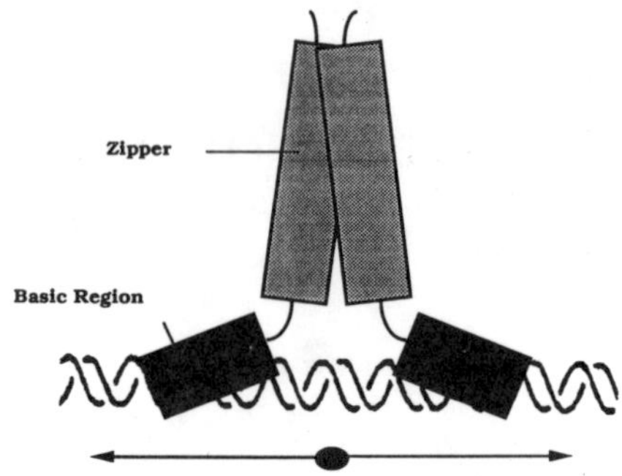

Fig. 1. A cartoon illustrating the general architecture of a bZIP protein bound to DNA. The light grey rectangles represent the alpha helices of the leucine zippers, which interact to form a short coiled-coil. This interaction positions the basic regions (dark grey rectangles) so that they can bind to the DNA. The overall two-fold rotational symmetry of the complex is indicated by the arrows.

correctly positioned for binding to the major groove of the DNA (Fig. 1). Dimerization is mediated by attractive interactions between the hydrophobic side chains of the leucine residues that are arranged in a "heptad repeat" that is the characteristic signature of the leucine zipper (*3*). For all members of the bZIP family, DNA binding is contingent upon dimerization, with monomers incapable of forming homodimers (such as those of the Fos family) (*4–6*) exhibiting no specific DNA binding. Many bZIP proteins exhibit the ability to form heterodimers, a property that raises interesting regulatory possibilities, many of which have been discussed elsewhere (*7, 8*). In this paper, we concentrate on specific regulatory questions that arise from the *in vitro* properties of one subfamily of bZIP proteins, the C/EBPs (CCAAT/enhancer binding proteins).

I. THE C/EBP FAMILY

C/EBP (now called C/EBPα) was initially identified as an activity in rat liver nuclear extracts capable of binding to both the CCAAT and enhancer core motifs (*9, 10*). Since the characterization of the gene for C/EBPα (*11*), four related proteins (C/EBPβ, γ, δ, and ε) have been identified. The genes encoding C/EBPs β (variously named LAP, AGP/EBP, NF-IL6, CRP2), δ (also called CRP3) and ε (also CRP1) were isolated by screening cDNA or genomic libraries at reduced stringency with DNA probes that encode the bZIP domain of C/EBPα (*12, 13*).

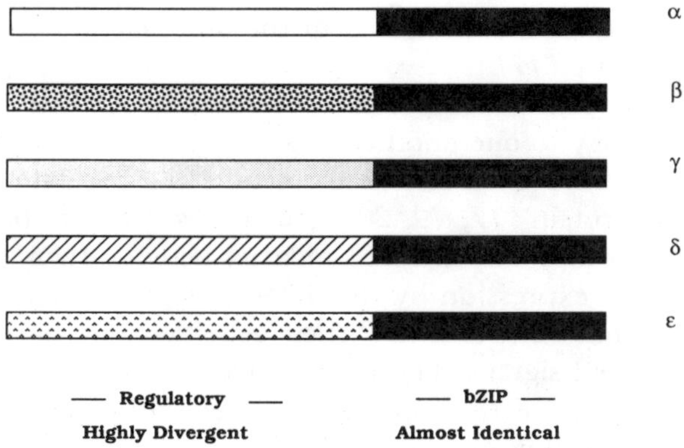

Fig. 2. A schematic diagram illustrating the homology between members of the C/EBP family. The regulatory regions, which represent about two thirds of the protein in each case, are highly divergent, as indicated by their different shadings. The bZIP (DNA binding) domains, which are divided into the basic region (black rectangle) and the leucine zipper (dark grey rectangle) are almost identical, and therefore have the same shading in each protein.

The gene for C/EBPγ (originally called Ig/EBP) was isolated by screening an expression cDNA library with a binding site from the immunoglobulin H gene enhancer (*14*). Several groups have also isolated C/EBPβ using similar methodology (*15–17*). Together with C/EBPα, these proteins form a distinct sub-family within the bZIP class. The assignment of these proteins into a distinct sub-family is based on two related observations. First, within the bZIP domain, these proteins are much more homologous to each other than to other bZIP proteins (Fig. 2) (*12–14*). This is especially true of the basic regions (the DNA contact surface) where a very high degree of homology exists. The leucine zippers are less highly related, but are still more similar to each other than to the zippers of other bZIP proteins. This similarity probably accounts for the fact that in all combinations tested to date, these proteins are capable of forming heterodimers (*12–14*). This ability to cross-dimerize is the second criterion for grouping these proteins into a sub-family. Using similar precepts, other bZIP proteins can be assigned to sub-families, such as those formed by the ATF/CREB proteins (*18*) and the Fos/Jun proteins (*6*).

The similarity of the bZIP domains may explain the remarkably similar DNA binding properties exhibited by C/EBP family members when assayed *in vitro*. Experiments employing gel shift or footprinting assays reveal that for the cases tested, C/EBP family members can bind to the same sites with similar affinities (*12*, *13*, *17*). For example, C/

EBPα and β bind to the same site in the albumin promoter with little difference in affinity (*17*).

Outside the bZIP domain, members of the C/EBP family show almost no homology to one another (Fig. 2) (*12–14*). For C/EBPs α and β, these regions have been shown to include the regulatory or effector domains of the protein (*17*, *19*, *20*). This is also likely to be true for the other C/EBP proteins. The lack of similarity between them suggests that regulation of gene expression by the different C/EBPs may be mediated by different mechanisms, and that their activities may be modulated in response to different signals. These differences are particularly interesting in light of the ability of these proteins to heterodimerize *in vitro*, which allows for the possibility of mixing different regulatory properties within the same heterodimeric factor. Examination of the expression patterns of the different C/EBP family members reveals that such mixing is plausible *in vivo*. C/EBPα protein is expressed at high levels in hepatocytes, adipocytes and in cells lining the gut (*21*). Although C/EBPβ RNA is widely expressed, high levels of protein are restricted to the liver and lung (*12*, *13*, *17*). Expression patterns for C/EBPδ, γ and ε proteins are currently unknown, although C/EBPδ and γ mRNAs are widely expressed (*12–14*). Thus the different C/EBP proteins show partially overlapping patterns of expression. Immunohistochemical studies confirm that the expression patterns of C/EBP family members can overlap at the cellular level (Z. Cao and S.L. McKnight, unpublished). For example, in adult rat liver hepatocytes, both C/EBPα and C/EBPβ are present in the nuclei of certain cells. Given the ability of these proteins to form heterodimers *in vitro*, it seems reasonable to suppose that they can also form heterodimers in hepatocyte nuclei.

1. Implications of the Properties of the C/EBP Family

These properties of the C/EBP family members raise a number of questions with respect to regulation of gene expression. Here, we will concentrate on the issue of selectivity. First, the DNA sequence specificity exhibited by C/EBPα is remarkably relaxed compared to that shown by prokaryotic regulatory proteins (*7*). This observation forces consideration of whether other factors contribute to selective binding of C/EBP proteins to enhancers and promoters. Second, the C/EBP family members show only small differences in affinity for a particular C/EBP site. In the case of the albumin promoter, it is known that C/EBPα and C/EBPβ bind to a site at −100 with respect to the transcriptional start site with similar affinities *in vitro* (*17*). Thus, in hepatocytes that have C/EBPα

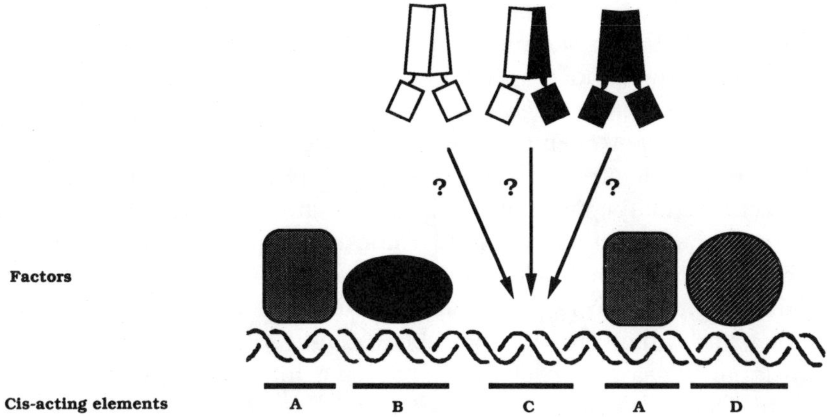

Fig. 3. Multiple C/EBP dimers can potentially bind to the same *cis*-regulatory sequence. In a cell expressing two C/EBP family members, three C/EBP dimers will be present, two homodimers and a heterodimer. These are illustrated at the top of the figure with black and white rectangles used to differentiate the two C/EBP proteins. The dimeric C/EBPs all have a similar affinity for a regulatory site, C, which is surrounded by other *cis*-elements, A, B, and D. These elements are the binding sites for other transcription factors which are depicted by the shaded figures.

homodimers, C/EBPβ homodimers and presumably C/EBPα-β heterodimers, all three forms could potentially bind to the same site (Fig. 3). For a promoter such as the albumin promoter, the question then arises as to whether there is any discrimination with respect to which C/EBP dimer actually binds *in vivo*. The highly divergent N-terminal regions of the various C/EBP family members, and the resultant potential for them to exert different effects on gene expression, suggests that discrimination with respect to which C/EBP dimer binds at a given site could be important. It seems reasonable to suggest, therefore, that some mechanism operates to provide a degree of selectivity over and above that provided by protein-DNA interactions. In the remainder of this paper we discuss a mechanism by which specificity may be achieved at promoters and enhancers, and describe attempts to develop an assay to detect such effects *in vitro*.

2. Interactions between DNA Bound Transcription Factors May Generate Specificity

Eukaryotic promoters and enhancers consist of a patchwork of *cis*-regulatory sequences that are the binding sites for transcription factors. As discussed above, each of these *cis*-acting sequences can potentially be bound by a variety of regulatory proteins. We (*7*, *22*) and others (*23*, *24*) have suggested that discrimination between transcription factors can be achieved not only through specific interactions with the DNA, but also by specific protein-protein interactions with factors bound at adjacent sites in the promoter. For example, the *cis*-regulatory sequences flanking a C/EBP site may differ from promoter to promoter, resulting in the binding of different regulatory proteins adjacent to the C/EBP site in each case. These neighboring proteins could impose a degree of selectivity with respect to which homo- or heterodimeric form of C/EBP would best "fit" that site (Fig. 4). Overall, the "fit" of different forms of C/EBP to a promoter or enhancer can then be thought of in terms of a three dimensional "jigsaw puzzle" (*22*), involving interactions with both protein and DNA. Within the C/EBP family, an excellent candidate for participation in such discriminatory interactions is the highly diverged N-terminal section of each protein. Thus the N-terminal region of C/EBPα could be envisaged to interact with a different set of transcription factors than that of C/EBPβ (Fig. 4). This would be achieved by the formation of attractive or repulsive molecular contacts between the C/EBP dimer and adjacent DNA-bound factors. These interactions are anticipated to be weak, and thus form only when the local concentration

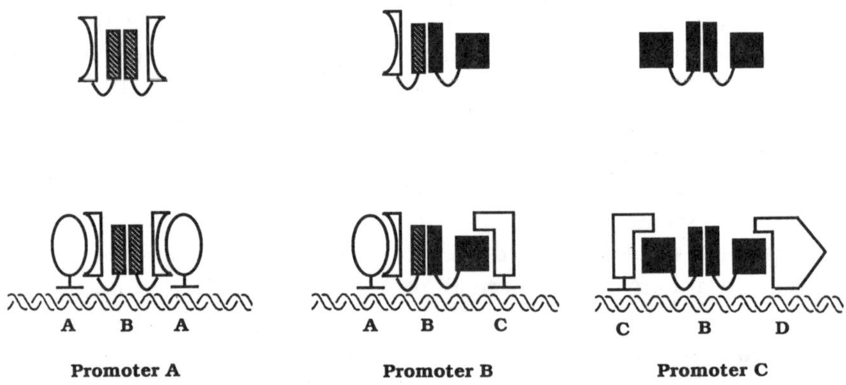

Fig. 4. Interactions between DNA-bound transcription factors. In this simplified scheme, three C/EBP dimers (two homodimers and a heterodimer) shown at the top of the figure, bind to a site B in three promoters with similar affinity. In each promoter, site B is flanked by different arrays of *cis*-regulatory sequences (labelled A, C, and D), which bind other proteins. Each C/EBP dimer is capable of forming a particular set of protein-protein contacts with other factors, as illustrated by the different shapes of their N-terminal regions. These contacts determine which promoter is the best "fit" for a given C/EBP dimer, as illustrated in the lower half of the figure.

of participating surfaces is high as a result of proteins binding to adjacent sites on the DNA template. Such template directed interactions have recently been described for the pituitary specific transcription factor Pit-1 (*25*). In order to test these ideas, we have developed a system in which to assess the contribution of different C/EBP family members to the transcription of several different promoters.

II. DEVELOPMENT OF A C/EBP DEPENDENT *IN VITRO* TRANSCRIPTION SYSTEM

Rat liver nuclear extracts prepared as described by Schibler and colleagues (*26*) are capable of accurately initiating transcription on a variety of promoters, and to some extent recapitulate the tissue specificity shown by some promoters *in vivo*. In initial experiments we have studied two promoters, the albumin promoter, which is highly active only in liver cells, and a synthetic promoter consisting of two strong C/EBP binding sites cloned immediately upstream of the TATA box of the Herpes Simplex virus thymidine kinase (TK) gene (Fig. 5). Mutation or deletion of the C/EBP sites in these promoters results in several fold decreases in the efficiency of transcription *in vitro* (*27*, P.L. and S.L.M., unpublished). Since the extracts contain both C/EBPα and C/EBPβ, we compared the activity of the two promoters in extracts specifically depleted for both C/EBPα and β with that seen in untreated extracts. Specific depletion for C/EBPα and β was achieved using immuno-affinity resins consisting of antibodies directed against C/EBPα and β covalently linked to agarose beads. Western blot analysis confirmed that depletion of the extract was successful. Both the albumin and synthetic promoters were transcribed several fold less efficiently in the α/β depleted extract than in the untreated extract. Thus depletion of C/EBP proteins has the same effect as mutation of the C/EBP sites, indicating that one (or more) of the C/EBP dimers in the extract is in fact responsible for activating expression from these promoters via these sites.

The depleted rat liver nuclear extract thus provides an excellent system for testing the effects of adding back exogenous C/EBPs. In an initial set of experiments, we have tested the effects of adding back a derivative of C/EBPα that has been expressed in bacteria and partially purified. The C/EBP derivative, called Δ1-2, lacks 58 amino acids near the amino terminus of the protein, and has been used in these experiments because it is efficiently expressed in bacteria, in contrast to full length C/EBPα. Transient transfection experiments have previously demonstrated that this deletion has little effect on the ability of C/EBPα to activate transcription (*19*). When added to the depleted rat liver nuclear extract, this protein efficiently activates transcription from the synthetic promoter (Fig. 5). However, it is unable to activate transcription in parallel experiments performed with the albumin promoter (Fig. 5). We are currently expressing C/EBPβ, in order to test whether in contrast to C/EBPα, it can activate the albumin promoter. If this proves to be the case we will be able to pinpoint the regions of the proteins responsible for their different activation potentials.

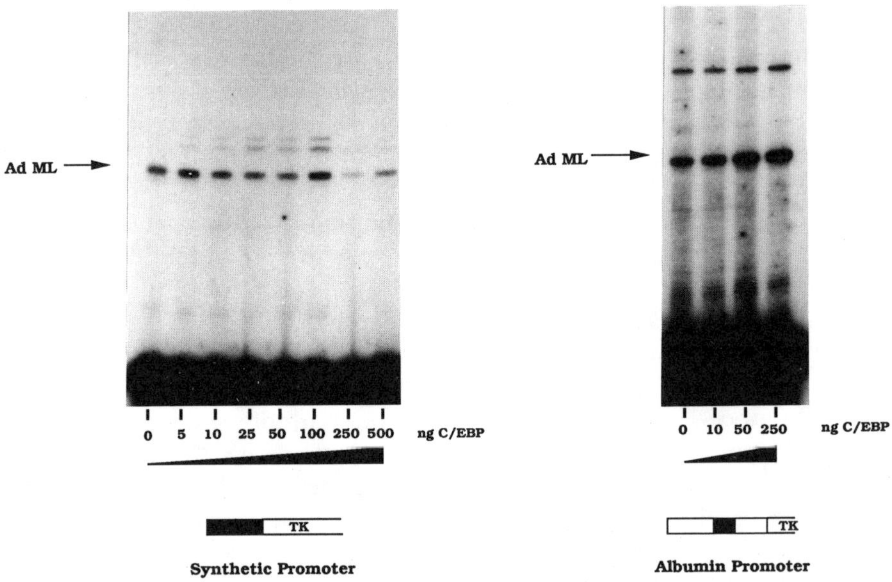

Fig. 5. Differential transactivation of promoters by recombinant C/EBPα *in vitro.* A derivative of C/EBPα, Δ1-2, was expressed in bacteria and partially purified. Increasing amounts of this protein were added to rat liver nuclear extracts depleted of endogenous C/EBPα and β. *In vitro* transcription reactions were then performed using either a synthetic promoter or the albumin promoter as a template. The location of C/EBP binding sites in each promoter is indicated by the stippled boxes in the schematic at the bottom of the figure. Each reaction also contained a control template consisting of the adenovirus major late promoter. The RNA synthesized in each reaction was purified and hybridized to a 3′ end labelled oligonucleotide complementary to the TK gene to which each promoter is fused. The hybridized oligo was then extended using reverse transcriptase, and the products separated on polyacrylamide gels. Autoradiographs of the dried gels are shown. The position of the extension product corresponding to the control major late promoter is indicated by the arrow labelled Ad ML.

SUMMARY

The bZIP family of DNA binding proteins utilize a bipartite DNA binding domain consisting of a basic region that contacts DNA linked to a leucine zipper that provides a dimerization interface. Many bZIP proteins have a limited ability to form heterodimers with one another, allowing for a mixing of the properties of each subunit. The C/EBP family of transcription factors constitutes such a cross-dimerizing sub-family of bZIP proteins. The five members of this family show a high degree of amino acid sequence identity in their bZIP domains, but have no similarity in the regulatory domains that constitute the N-terminal two thirds of each protein. The *in vivo* expression patterns of the C/EBPs overlap, such that some cells will contain multiple C/EBP dimers. The *in vitro* DNA binding properties of the different dimers raises interesting questions about how specific regulation of gene expression is achieved by the C/EBP family. Members of the C/EBP family show a relaxed specificity for DNA sequence, and the different homodimers exhibit similar affinities for individual C/EBP sites. We therefore suggest that specific contacts between C/EBPs and other DNA bound transcription factors have an important role in generating specificity with respect to which dimeric C/EBP factor will bind a particular promoter or enhancer. Discrimination between different C/EBP dimers could be achieved if the divergent N-terminal sections are capable of participating in different sets of protein-protein contacts. Thus the "fit" of a given C/EBP dimer to a promoter or enhancer depends not only on interactions with the DNA, but also on the complementarity of other regions of the protein to adjacent DNA bound factors, rather like the fit of an individual piece into a jigsaw puzzle.

In order to investigate the degree of specificity of promoter regulation exhibited by members of the C/EBP family, we have developed an *in vitro* transcription system that is responsive to exogenously supplied C/EBP. We have shown that in this system, C/EBPα shows a context dependent ability to activate transcription. These studies can now be extended to other members of the C/EBP family, and should enable us to determine the molecular basis for selective regulation of gene expression within the C/EBP family.

Acknowledgments

We thank Zhoadan Cao for permission to site unpublished work, and C. Thompson for comments on the manuscript.

REFERENCES

1. C.R. Vinson, P.B. Sigler, and S.L. McKnight: *Science* **246**, 911–916 (1989).
2. W.H. Landschulz, P.F. Johnson, and S.L. McKnight: *Science* **243**, 1681–1688 (1989).
3. W.H. Landschulz, P.F. Johnson, and S.L. McKnight: *Science* **240**, 1759–1764 (1988).
4. T.D. Halazonetis, K. Georgopoulos, M.E. Greenberg, and P. Leder: *Cell* **55**, 917–924 (1988).
5. Y. Nakabeppu, K. Ryder, and D. Nathans: *Cell* **55**, 907–915 (1988).
6. C. Abate and T. Curran: *Semn. Cancer Biol.* **1**, 19–26 (1990).
7. P. Lamb and S.L. McKnight: *Trends Biochem. Sci.* **16**, 417–422 (1991).
8. N. Jones: *Cell* **61**, 9–11 (1990).
9. B.J. Graves, P.F. Johnson, and S.L. McKnight: *Cell* **44**, 565–576 (1986).
10. P.F. Johnson, W.H. Landschulz, B.J. Graves, and S.L. McKnight: *Genes Dev.* **1**, 133–146 (1987).
11. W.H. Landschulz, P.F. Johnson, E.Y. Adashi, B.J. Graves, and S.L. McKnight: *Genes Dev.* **2**, 786–800 (1988).
12. Z. Cao, R.M. Umek, and S.L. McKnight: *Genes Dev.* **5**, 1538–1552 (1991).
13. S.C. Williams, C.A. Cantwell, and P.F. Johnson: *Genes Dev.* **5**, 1553–1567 (1991).
14. C. Roman, J.S. Platero, J. Shuman, and K. Calame: *Genes Dev.* **4**, 1404–1415 (1990).
15. S. Akira, H. Isshiki, T. Sugita, O. Tanabe, S. Kinoshita, Y. Nishio, T. Nakajima, T. Hirano, and T. Kishimoto: *EMBO J.* **9**, 1897–1906 (1990).
16. C-J. Chang, T-T. Chen, H-Y. Lei, D-S. Chen, and S-C. Lee: *Mol. Cell. Biol.* **10**, 6642–6653 (1990).
17. P. Descombes, M. Chojkier, S. Lichsteiner, E. Falvey, and U. Schibler: *Genes Dev.* **4**, 1541–1551 (1990).
18. T. Hai, F. Liu, W.J. Coukos, and M.R. Green: *Genes Dev.* **3**, 2083–2090 (1989).
19. A.D. Friedman and S.L. McKnight: *Genes Dev.* **4**, 1416–1426 (1990).
20. D. Pei and C. Shih: *Mol. Cell. Biol.* **11**, 1480–1487 (1991).
21. E.H. Birkenmeier, B. Gwynn, S. Howard, J. Jerry, J.I. Gordon, W.H Landschulz, and S.L. McKnight: *Genes Dev.* **3**, 1146–1156 (1989).
22. P.F. Johnson and S.L. McKnight: *Annu. Rev. Biochem.* **58**, 799–839 (1989).
23. S. Hayashi and M.P. Scott: *Cell* **63**, 883–894 (1990).
24. A.D. Frankel and P.S. Kim: *Cell* **65**, 717–719 (1991).
25. H.A. Ingraham, S.E. Flynn, J.W. Voss, V.R. Albert, M.S. Kapiloff, L. Wilson, and M.G. Rosenfeld: *Cell* **61**, 1021–1033 (1990).
26. K. Gorski, M. Carneiro, and U. Schibler: *Cell* **47**, 767–776 (1986).
27. P. Maire, J. Wuarin, and U. Schibler: *Science* **244**, 343–346 (1989).

Molecular Structure and Life (Ed. by Y. Kyogoku and Y. Nishimura), pp. 183–194, Japan Sci. Soc. Press, Tokyo/CRC Press, Boca Raton (1992)

The Crystal Structure and RNA Binding Properties of the RNP Domain of the U1 A Small Nuclear Ribonucleoprotein

Chris OUBRIDGE, Timm-H. JESSEN, Jade LI, C. Hiang TEO, Clare PRITCHARD, Phil R. EVANS, and Kiyoshi NAGAI

*MRC Laboratory of Molecular Biology**

I. U1 SMALL NUCLEAR RIBONUCLEOPROTEIN PARTICLE (U1 snRNP)

In eukaryotic genes protein coding sequences are interrupted by sequences of unknown function (introns). In the cell nucleus the entire length of the gene, both introns and exons, is transcribed into RNA. This RNA (pre-messenger RNA) is then processed and the introns are removed. Several small nuclear ribonucleoprotein particles (snRNPs or 'snurps') take part in the excision of introns from pre-messenger RNAs (*1*). These snRNPs are large protein-RNA complexes consisting of small nuclear RNAs (snRNAs) and many protein subunits (*2*). U1 and U2 snRNP recognise the 5′ splice site and branch point of pre-messenger RNA, respectively (*3*). U1 snRNP consists of a single RNA chain of 165 nucleotides (U1 RNA), three U1 specific subunits: A, 70k and C proteins and several core or Sm proteins which also form a core structure in U2, U5, and U4 snRNPs (*2*). Phylogenetic and chemical footprinting data show that U1 RNA folds into a secondary structure with four stem/loop

* Cambridge CB2 2QH, U.K.

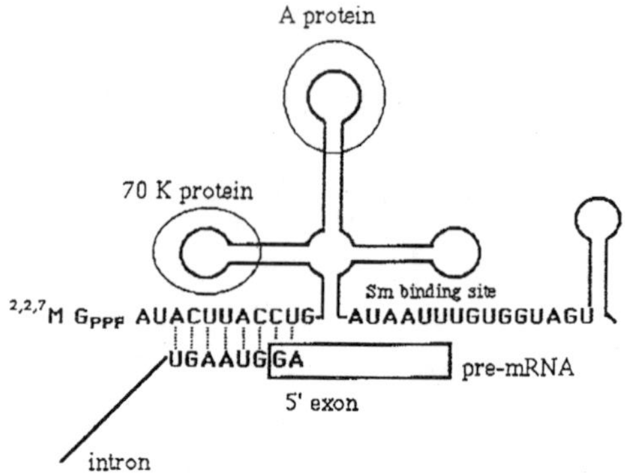

Fig. 1. Structure of U1 snRNP showing the base-pairing interaction with the 5′ splice site of pre-mRNA.

structures, as shown in Fig. 1 (*4*). Biochemical data show that the U1 70k and U1 A proteins bind to stem/loop I and II, respectively (*5*). Core proteins bind to a single-stranded U-rich region flanked by two stem/loop structures. U2, U5, and U4 RNAs also have similar U-rich sequences flanked by stem/loop structures and this common feature is recognised by the core proteins. The 5′ end of U1 RNA is complementary to the conserved sequence at the 5′ splice site and genetic evidence suggests that base-pairing between the 5′ end of U1 RNA and the 5′ splice site plays an important role in 5′ splice site recognition (*6*). In order to understand this important cellular process it is essential to elucidate the structure of snRNPs.

II. THE U1 A PROTEIN CONTAINS RNP DOMAINS

The amino acid sequence of the U1 A protein shows that it consists of two copies of the RNP (ribonucleoprotein) domain connected by a linker peptide. The RNP domain was first found in yeast poly-A binding protein as a domain repeated four times in tandem (*7*). Homologous domains have now been found in many proteins of diverse function. The RNP domain is now considered to be an RNA binding module which appeared early in evolution and has given rise to many RNA binding proteins with highly diverse functions. The present RNP domain-containing proteins are thought to have arisen from gene duplication events and recombination with other protein-structural domains. The RNP domain consists of about 90 residues and contains two highly conserved short segments, normally referred to as RNP1 and RNP2. Proteins containing the RNP domain are distributed in a variety of organisms from bacteria to man and are involved in important biological processes such as transcriptional termination, pre-mRNA splicing, RNA transport, and mRNA translation. We have solved the structure of the RNP domain of the U1 A protein and studied its RNA binding properties to understand how this domain interacts with its cognate RNA hairpin (*8*, *9*).

III. THE CRYSTAL STRUCTURE OF THE RNP DOMAIN OF THE U1 A PROTEIN

The N-terminal RNP domain of the U1 A protein, containing residues 1–95 (A95 protein), was produced in *Escherichia coli* using the T7 RNA polymerase expression vector. Crystals of the A95 protein were grown from about 2 M Na/K phosphate buffer pH 5.5 by micro-dialysis. These crystals belong to the cubic space group $I4_132$ with unit cell edges of 148.1 Å. Ten mutant proteins were engineered where single polar side chains were replaced by cysteines. Methylmercury derivatives of four of the mutants produced crystals isomorphous with the native crystals. The structure was solved by conventional isomorphous replacement methods, combined with solvent flattening, and the map was readily interpretable. The asymmetric unit contains two molecules of the A95 protein and the initial model building was carried out using the program O (*10*); the polypeptide chain was traced with the Bones option, using the heavy atom sites as markers for their respective residues, and the main chain and side chains were built to the trace using the auto-build options. After some adjustment of side chain conformations the model was refined by simulated annealing with X-plor (*11*), which reduced the R factor from 0.44 to 0.28 for all data from 10 to 2.8 Å resolution, without refinement of temperature factors. Further refinement by the method of Hendrickson and Konnert (*12*) with individual atom temperature factors, and some minor rebuilding, reduced the R factor to 0.22 (see Table I). Refinement at 2.4 Å resolution is now in progress (Evans, Oubridge, and Nagai, unpublished results). Figure 2 shows a schematic representation of the RNP domain of the U1 A protein. The RNP domain consists of a four-stranded anti-parallel β sheet, flanked on one side by two α helices. These secondary structural elements are arranged in the order of β1-A helix-β2-β3-B helix-β4 in the primary structure. The conserved RNP1 and RNP2 segments are located in the two central β strands (*8*). The structure of the two molecules in the asymmetric unit are identical except

TABLE I
Crystal Data

Derivatives	Resolution (Å)	No. of unique reflections	Mean redundancy	R_{merge}	No. of heavy atom sites	$\langle F_H \rangle / E$
Native	2.8	7,067	7.7	0.090	—	—
S71C+MeHg	3.5	3,509	5.0	0.101	4	1.7
Q85C+MeHg	2.8	6,913	8.0	0.082	4	1.2
Q39C+MeHg	2.8	6,674	7.2	0.077	2	1.4
E61C+MeHg	3.0	5,774	7.7	0.087	2	1.3

$R_{merge} = \Sigma | Ij - \langle I \rangle | / \Sigma \langle I \rangle$ where Ij is an individual measurement of intensity and $\langle I \rangle$ is the mean intensity for this reflection. $\langle F_H \rangle / E$, phasing power, (r.m.s. calculated F_H)/(r.m.s. error), for centric reflections only.

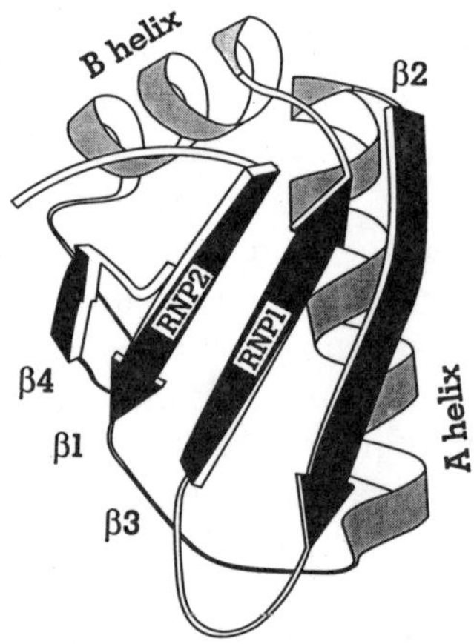

Fig. 2. Schematic ribbon representation of the RNP domain of the U1 A protein elucidated by x-ray crystallography. The RNP domain consists of a four-stranded β sheet and two α helices. The RNP1 and RNP2 segments lie on the $\beta 3$ and $\beta 1$ strands, respectively.

at both termini and in the loops between the $\beta 1$ strand and the A helix and between the $\beta 2$ and $\beta 3$ strands. This implies that these regions are relatively flexible. In these two loop regions there are many basic residues (K22, K23, K27, R47, K50, and R52) and these two loops form a pair of basic 'jaws.'

IV. RNA BINDING STUDIES OF THE U1 A PROTEIN

In order to map the RNA binding surface of the U1 A protein we have studied the RNA binding properties of mutant proteins. Scherly *et al.* (*13*) showed that the fragment of the U1 A protein containing the first

102 residues has full sequence specific RNA binding activity and therefore we have studied U1 RNA binding properties of mutants of a protein containing the first 102 residues, A102. This fragment contains 11 Lys and 5 Arg residues, some of which are likely to interact with the backbone phosphates or bases of the RNA. We replaced each Lys and Arg residue in turn by Gln and determined the RNA binding affinity of each mutant (*8*). The specific RNA binding was abolished in the R52Q mutant and a 10–50 fold reduction in binding was observed with the K22Q, K27Q, K50Q, K80Q, and K98Q mutants.

We then mutated residues on the four-stranded β sheet and studied the RNA binding of the resultant proteins (*9*). By examining the crystal structure we chose substitutions in such a way that minimal instability and structural disturbance was caused in the protein. Figure 3 shows the surface of the four-stranded β sheet; Table II summarises RNA-protein dissociation constants of the mutants on the RNA binding surface. The conserved RNP2 segment comprises the β1 strand and the β1-A helix loop region. Ile12 and Ile14 are internal residues which form part of the protein core, whereas Thr11, Tyr 13, Asn15, and Asn16 are on the

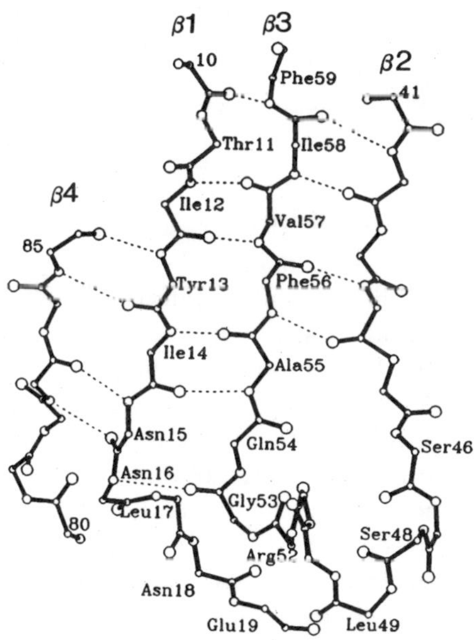

Fig. 3. Hydrogen bonding network of the four stranded β sheet. The conserved RNP1 (residues 12–17) and RNP2 (residues 52–59) lie side by side in the middle two β strands.

TABLE II
Relative Binding Affinities of Mutant A102 Proteins to U1 RNA Stem/loop II

Protein	K_D [nM]	Residue in U2 B″ protein	Location within structure
A102 wt	10		
A96	60		
N 9 V	40	N	N-terminal loop
H 10 N	10	H	″
T 11 V	—	T	β1 (RNP2)
Y 13 F	250	Y	″
N 15 V	—	N	″
N 16 V	150	N	″
N 18 A	50	N	Loop β1-helix A
E 19 D	20	D	″
Q 39 H	10	H	Loop helix A-β2
S 46 A	20	L	β2
S 46 L	90	L	″
S 48 A	20	T	Loop β2-β3
S 48 T	10	T	″
S46LS48T	80	L,T	β2, loop β2-β3
L 49 A	20	M	Loop β2-β3
R 52 Q	—	R	″ (RNP 1)
R 52 K	30	R	″ ″
Q 54 F	1,400	Q	β3 ″
F 56 Y	30	F	″ ″
D 79 V	30	G	Loop helix B-β4
Q 85 V	20	Q	β4

The dissociation constant K_D of the U1 RNA stem/loop II with various mutant proteins is shown in comparison with the wild-type (wt) A102 protein (amino acids 1-102 of the U1 A protein). Total loss of specific binding is indicated by —; K_D was reproducible within $\pm 20\%$.

surface of the β sheet and could interact with the RNA. In the crystal structure, the hydroxyl group of Thr11 forms no hydrogen bond with other residues and the Thr11→Val mutation is therefore unlikely to cause any significant structural changes. However the mutation completely abolished U1 RNA binding, indicating that this residue makes crucial RNA contacts. The Asn15→Val mutation also completely abolished the U1 RNA binding and the Asn16→Val mutation also substantially reduced the RNA binding. The phenol oxygen of Tyr13 forms a hydrogen bond with the side chain of Gln54 and the Tyr13→Phe mutation substantially reduced the U1 RNA binding. This suggests that the phenol oxygen may either form a hydrogen bond directly with RNA or orientate the Gln54 side chain to permit formation of a hydrogen bond with RNA.

The RNP1 segment comprises the β3 strand and its preceding loop (β2–β3). The Arg52→Gln mutation completely abolishes the RNA binding whereas the Arg52→Lys mutation only slightly reduced the RNA binding. This residue may be involved in contacts with the phosphate backbone. The RNA binding is substantially reduced by the Gln54→Phe mutation. Our mutagenesis experiments show that U1 RNA loop II binds to the surface of the four-stranded β sheet and to loops at one edge of the β sheet.

V. RNA BINDING SPECIFICITIES OF THE U1 A AND U2 B″ PROTEINS

The U2 B″ protein is a component of U2 snRNP that binds to stem/loop IV of U2 RNA in the presence of the U2 A′ (Fig. 4 and refs. *14*, *15*). The U2 B″ protein also consists of two RNP domains (*16*) of which only the N-terminal RNP domain is needed to form a tight complex with the U2 A′ protein, and to confer full binding specificity to U2 snRNA (*14*, *15*). The U1 A and U2 B″ proteins show about 90% sequence identity in their N-terminal domains (*17*). Residues variable between these two proteins are dispersed throughout the domain, but there is a cluster of five residues (VSRSL) in the β2 strand of the U1 A protein which is replaced by ALKTM in the U2 B″ protein. When these five residues in U2 B″ protein are replaced by those found in the U1 A the resulting chimera bound to U1 RNA stem/loop II (*14*). This experiment suggests that these five residues play an important role in distinguishing between the two closely related RNA stem/loops.

The nucleotide sequence of the stem/loop IV of U2 RNA, the binding site for the U2 B″ protein, is very similar to that of the stem/loop II of U1 RNA (Fig. 4). By swapping the loop sequences of these two RNAs, Scherly *et al.*(*14*) showed that differences in the protein binding specificity are principally attributable to the nucleotide sequence in the loops. Since the 5′ end of the loop sequence (AUUGCA) is conserved in both stem/loop II and IV sequences it seems likely that the 5′ end of the loop interacts with residues in the RNP1 and RNP2 segments, where amino acid residues are conserved. The variable 3′ ends of the two RNA loops might be expected to interact mainly with the β2 segment where the protein sequences also differ. These considerations restrict the possible orientation of the loop on the β sheet to one of two orientations, depending on whether the minor or major groove side of the loop interacts with the β sheet, as shown in Fig. 5. For both models I and II the conserved AUUGCA sequence is in contact with the RNP1 and RNP2 segments and the variable 3′ half of the loop is in contact with the

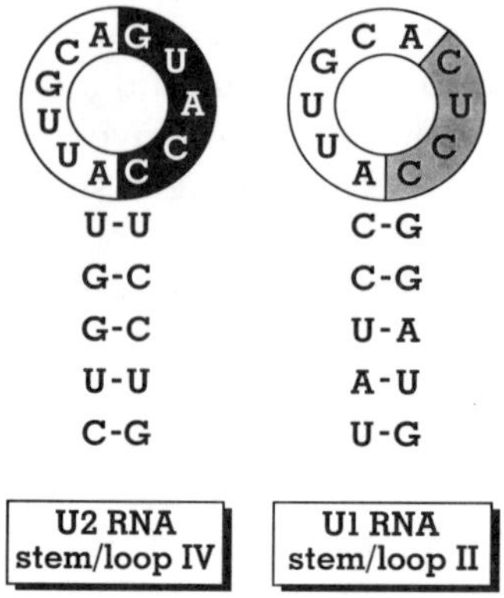

Fig. 4. Sequences of the stem/loop IV of U2 snRNA and stem/loop II of U1 snRNA. The shaded regions indicate where loop sequences differ.

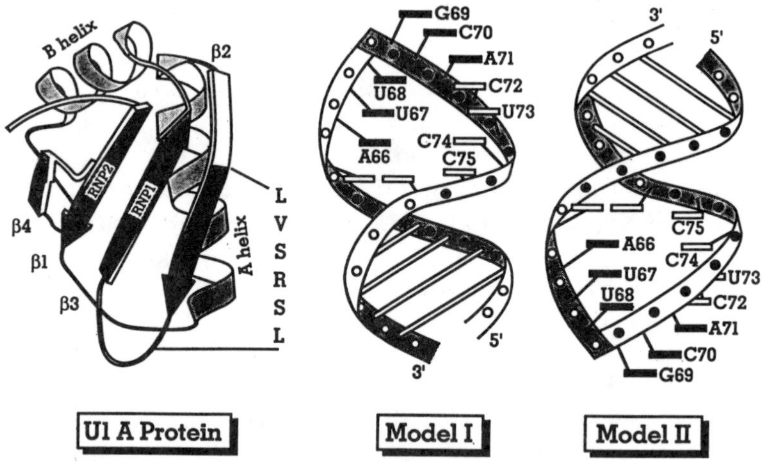

Fig. 5. Shows schematic representations of the N-terminal RNP domain of U1 A protein and the stem/loop II of U1 snRNA (not to scale). The two orientations of the stem/loop RNA correspond to the two models of RNA binding to protein which allow interaction between the 3′ (variable) region of the loop and the unconserved section of the $\beta2$ strand of the protein (filled in black). Phosphates protected from ethylation when RNA is bound to protein are shown as filled circles. The conserved bases AUUGCA are represented by black bars.

$\beta2$ strand. Ethylation protection studies were undertaken to distinguish between these two possibilities (*9*). RNA backbone phosphate groups are uniformly ethylated by ethylnitrosourea, whereas the backbone phosphates of the 5′ stem and the 3′ two thirds of the loop are clearly protected in the presence of the U1 A protein (phosphate groups protected from ethylation are shown by black dots in Fig. 5). In model II the phosphates in the 5′ stem are facing away from the protein surface and this model cannot account for the protection from ethylation observed in these regions. In model I the phosphates in the 5′ stem are wedged between the pair of "basic jaws" and protection of phosphates in this region is well accounted for.

Krol *et al.* (*18*) proposed a model of U1 RNA based on chemical protection experiments. As in the anticodon loop of tRNA, which has seven unpaired bases, the loop II of U1 RNA is stabilised mainly by base

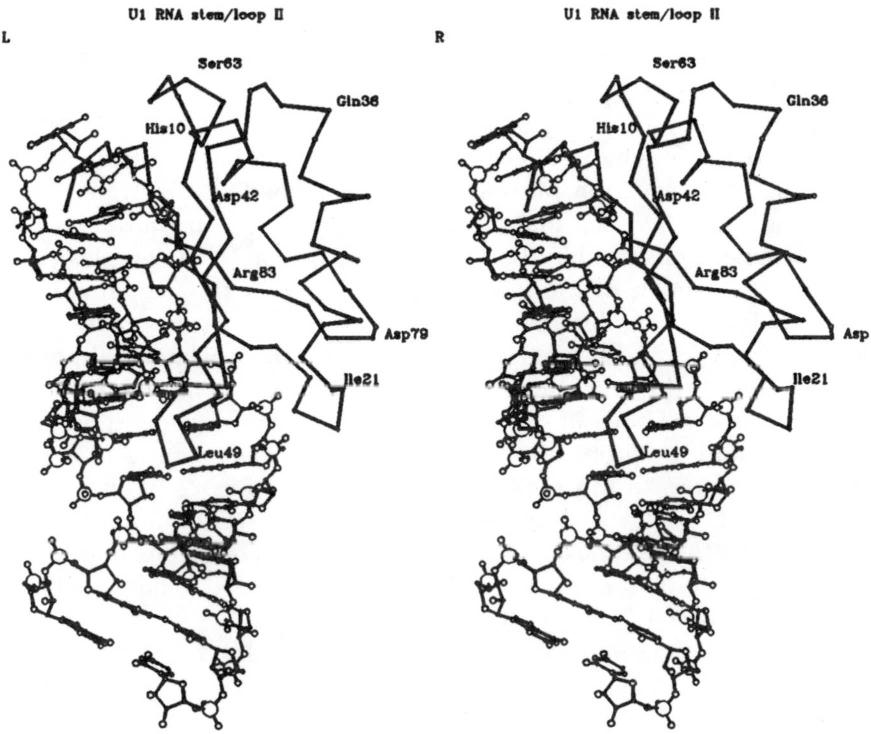

Fig. 6. Model of the complex of the N-terminal RNP domain of the U1 A protein and its cognate RNA in stereo. The Ca-backbone of the RNP domain (*8*) is represented as a solid line. All atoms of the U1 RNA stem/loop II (nucleotides 57–84) are shown according to the model proposed by Krol *et al.* (*18*). The view shows the interface between the four-stranded β sheet and the U1 RNA loop II. Some Cα atoms of the RNP domain are labelled. The 5′ stem of the U1 RNA loop II is embedded between a pair of basic jaws formed between the β1-A helix and β2–β3 loops.

stacking. In this model the phosphate backbone assumes a sharp turn in the phosphate diester bond between U68 and G69; bases are stacked on top of each other before and after this turn. We propose a model of the U1 A protein-RNA complex structure based on the crystal structure of the U1 A protein and the model of U1 RNA proposed by Krol *et al.* (*18*). This model, shown in Fig. 6, is consistent with the biochemical data discussed above and has some attractive structural features. As first pointed out by Chothia, the surface of an anti-parallel β sheet has a right-handed twist when viewed along the β strand (*19*). It therefore shows good surface complementarity to the surface of the U1 RNA loop which also has a right-handed twist. In the 3′ two thirds of the loop, phosphates point towards the surface of the β sheet where they could interact with the β2 strand and the loop between the β2 and β3 strands. This fits in well with the observations of Carter and Kraut (*20*) and Church *et al.* (*21*) who pointed out that a repeating unit of DNA or RNA phosphate backbone nearly coincides with that of an antiparallel β strand. The main chain amide groups of the β2 strand can form hydrogen bonds with the phosphates in the 3′ half of the RNA loop. Hydrogen bonds between main chain amide groups and phosphates of DNA have been observed in the 434 phage repressor-DNA complex (*22*). Ethylation protection of the phosphate backbone in this region can be accounted for by these hydrogen bonds. In the 5′ stem, phosphates lying between 5′ to C59 and 5′ to C64 are protected from ethylation, although the protection of the first and last phosphates is very weak. In our model the phosphate backbone in this region is wedged between the upper and lower loops and the protection from ethylation of this region is therefore well accounted for. In this region, the Lys 22→Gln, Lys 23→Gln, Lys 27 →Gln, Arg 47→Gln, Lys 50→Gln, and Arg 52→Gln mutations all reduce U1 RNA binding by various factors (*8*), suggesting that these residues hold the 5′ stem between the "basic jaws". The details of the interaction between the U1 A protein and U1 RNA remain to be determined by X-ray analysis of a crystalline complex.

SUMMARY

The U1 A protein is a component of U1 snRNP, which is involved in recognition of the 5′ splice site of pre-messenger RNA in the initial stages of pre-messenger RNA splicing. The U1 A protein binds to stem/loop II of U1 snRNA. The primary structure of U1 A protein shows that it contains two copies of the RNP domain separated by a linker peptide.

The RNP domain is a conserved sequence motif of around 90 amino acids found in a number of RNA-binding proteins. Biochemical experiments have shown that only the N-terminal RNP domain of U1 A protein is required for specific binding to U1 stem/loop II RNA.

We have solved the crystal structure of the N-terminal RNP domain of U1 A protein and shown that it consists of a four-stranded antiparallel-β sheet flanked on one side by two α helices. Footprinting experiments and extensive mutagenesis of the RNP domain, in conjunction with phylogenetic considerations, have suggested to us a model of the U1 A protein's interaction with its cognate RNA. In this model the major groove side of the RNA loop binds to β sheet; the RNA stem interacts with the flexible and basic loop regions of the protein, these are the "basic jaws" which lie at one edge of the β sheet.

Acknowledgments

We thank Dr. Max Perutz, Prof. Aaron Klug, Wes Sundquist, and Richard Turner for helpful discussion and criticism and Annette Lenton and Yuko Nagai for drawing the figures.

REFERENCES

1. J.A. Steitz, D.L. Black, V. Gerke, K.A. Parker, A. Kramer, D. Frendewey, and W. Keller: In "Structure and Function of Major and Minor snRNPs" (Ed. by M.L. Birnstiel), pp. 115–154, Springer-Verlag, New York (1988).
2. R. Lührmann, B. Kastner, and M. Bach: *Biochim. Biophys. Acta* **1087**, 265–292 (1990).
3. I.W. Mattaj: *Curr. Opin. Cell. Biol.* **2**, 528–538 (1990).
4. C. Guthrie and B.A. Patterson: *Annu. Rev. Genet.* **22**, 387–419 (1988).
5. J.R. Patton and T. Pederson: *Proc. Natl. Acad. Sci. U.S.A.* **85**, 747–751 (1988).
6. M. Rosbash and B. Seraphin: *Trends Biochem. Sci.* **16**, 187–190 (1991).
7. R.J. Bandziulis, M.S. Swanson, and G. Dreyfuss: *Genes Dev.* **3**, 431–437 (1989).
8. K. Nagai, C. Oubridge, T.-H. Jessen, J. Li, and P.R. Evans: *Nature* **348**, 515–520 (1990).
9. T-H. Jessen, C. Oubridge, C.H. Teo, C. Pritchard, and K. Nagai: *EMBO J.* **10**, 3447–3456 (1991).
10. T.A. Jones, J.Y. Zou, S.W. Cowan, and M. Kjeldegaard: *Acta Crystallogr.* **A47**, 110–119 (1991).
11. A.T. Brünger: *J. Mol. Biol.* **203**, 803–816 (1988).
12. W.A. Hendrickson and J.H. Konnert: In "Structure, Conformation and Evolution" (Ed. by R. Srinivasan), Vol. 1, pp. 43–57, Pergamon, Oxford (1981).
13. D. Scherly, W. Boelens, W.J. van Venrooij, N.A. Dathan, J. Hamm, and I.W. Mattaj: *EMBO J.* **8**, 4163–4170 (1989).
14. D. Scherly, W. Boelens, N.A. Dathan, W.J. van Venrooij, and I.W. Mattaj: *Nature* **345**, 502–506 (1990).

15. D. Scherly, N.A. Dathan, W. Boelens, W.J. van Venrooij, and I.W. Mattaj: *EMBO J.* **9**, 3675–3681 (1990).
16. W.J. Habets, P.T.G. Sillekens, M.H. Hoet, J.A. Schalken, A.J.M. Roebroek, J.A.M. Leunissen, W.J.M. van de Ven, and W.J. van Venrooij: *Proc. Natl. Acad. Sci. U.S.A.* **84**, 2421–2425 (1987).
17. P.T.G. Sillekens, W.J. Habets, R.P. Beijer, and W.J. van Venrooij: *EMBO J.* **6**, 3841–3848 (1987).
18. A. Krol, E. Westhof, M. Bach, R. Lührmann, J.-P. Ebel, and P. Carbon: *Nucleic Acids Res.* **18**, 3803–3811 (1990).
19. C. Chothia: *Annu. Rev. Biochem.* **55**, 537–572 (1984).
20. C.W. Carter and J. Kraut: *Proc. Natl. Acad. Sci. U.S.A.* **71**, 283–287 (1974).
21. G.M. Church, J.L. Sussman, and S.H. Kim: *Proc. Natl. Acad. Sci. U.S.A.* **74**, 1458–1462 (1977).
22. A.K. Aggarwal, D.W. Rodgers, M. Drottar, M. Ptashne, and S.C. Harrison, *Science* **242**, 899–907 (1988).

Molecular Structure and Life (Ed. by Y. Kyogoku and Y. Nishimura), pp. 195–205, Japan Sci. Soc. Press, Tokyo/CRC Press, Boca Raton (1992)

Molecular Communication between *Escherichia coli* RNA Polymerase and Transcription Factors

Akira ISHIHAMA

*Department of Molecular Genetics, National Institute of Genetics**

The RNA polymerase in prokaryotes is composed of a core enzyme with the subunit structure $\alpha_2\beta\beta'$ and one of the multiple molecular species of σ subunit. Polymerization of RNA chains is catalyzed by the core enzyme, but the σ subunit is necessary for maturation of this core enzyme and transcription initiation from the promoter DNA (reviewed in refs. *1*, *2*). Interconversion of a single species of RNA polymerase between different forms of transcription apparatus with different promoter selectivity is considered to be the major mechanism of global control of transcription (for review see refs. *3*, *4*). Two hierarchies of the reversible molecular assembly are involved in this functional interconversion of RNA polymerase, *i.e.*, replacement of the σ subunit leading to the generation of multiple forms of the holoenzyme, and reversible interaction of the RNA polymerase with one of the accessory transcription factors. In this report, I will summarize our studies on the molecular communication between RNA polymerase core enzyme, σ subunits, and transcription factors from *Escherichia coli*.

* Mishima, Shizuoka 411, Japan

I. COMMUNICATION BETWEEN RNA POLYMERASE CORE ENZYME AND σ SUBUNIT

1. *Subunit-subunit Contact within the Core Enzyme*

Core enzyme subunits of *E. coli* RNA polymerase are assembled sequentially in the order: $2\alpha \rightarrow \alpha_2 \rightarrow \alpha_2\beta \rightarrow \alpha_2\beta\beta'$ (inactive premature core) $\rightarrow$E (active core) (reviewed in ref. *5*). Subunits β and β' do not interact with each other, but subunit α plays an essential role in linking these two large subunits together. This immediately suggested that there is a contact site(s) on α with β (Fig. 1).

Since the α subunit is not directly involved in the catalytic function of RNA polymerization, only a limited number of *rpoA* mutants have been isolated. Only two temperature-sensitive (ts) *rpoA* mutations have been isolated to date (*6*). Sequence analysis indicated that these two ts mutations, *rpoA*112 and *rpoA*101, led to Arg to Cys substitution at the amino acid residues 45 and 191, respectively (*7*). In *E. coli* cells carrying the *rpoA*112 mutation, the RNA polymerase assembly is blocked at the step of either α subunit dimerization or $\alpha_2\beta$ complex formation (*8*). These observations suggested that either the α-α or the α-β contact site was located near the N-terminal region of α. Analysis of a set of *rpoA* mutants carrying various extents of deletion from the C-terminus indeed

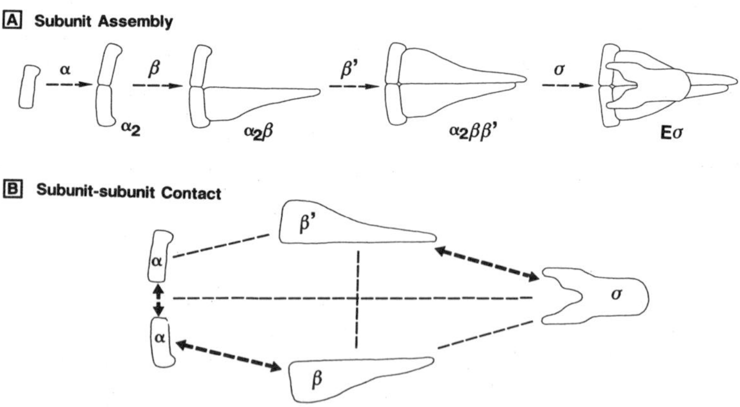

Fig. 1. Subunit-subunit contacts between the RNA polymerase subunits. (A) *E. coli* RNA polymerase holoenzyme is assembled both *in vitro* and *in vivo* sequentially in the indicated order. The inactive premature core enzyme ($\alpha_2\beta\beta'$) is converted into active core enzyme (E) following incubation in the presence of σ subunit. (B) Isolated subunits form α-α, α-β, and β'-σ complexes. Genetic studies indicate subunit-subunit interaction of other combinations, which may take place only within the assembled RNA polymerase.

indicated that the N-terminal proximal region of α subunit plays an essential role in the core enzyme assembly (*9, 10*) (Fig. 2).

On the other hand, temperature-sensitive mutations in the β (*rpoB*) or β' (*rpoC*) often interfere with the RNA polymerase assembly at non-permissive temperatures rather than the function of assembled RNA polymerase, suggesting that wide regions of the β and β' polypeptides are involved in the subunit-subunit contact. Glass and colleagues made a large collection of *rpoB* mutants, each carrying a single amino acid substitution on the β subunit (*11*). Using this mutant collection, we carried out a systematic mapping of functional sites on the β polypeptide, and determined that the C-terminal region of β was needed for the assembly of core enzyme (*12*) (Fig. 2). On the assembly pathway, subunit β' binds to $\alpha_2\beta$ complex to form $\alpha_2\beta\beta'$ premature core enzyme. Since neither α nor β forms a stable complex with β', it appears that the contact site to β' on one of these subunits is buried in the isolated state

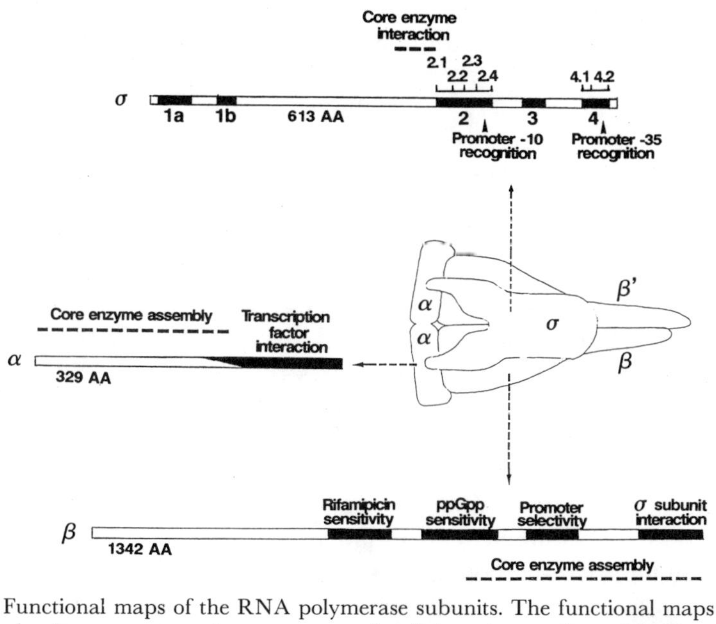

Fig. 2. Functional maps of the RNA polymerase subunits. The functional maps of the α and β subunits are a tentative summary of collaborative studies with Hayward and colleagues (Edinburgh) and Glass and colleagues (Nottingham), respectively. The map of the σ^{70} subunit shows the location of sequences which are conserved among σ family proteins from diverse prokaryotes. Genetic studies indicated the regions for core enzyme binding, promoter −10 recognition, and promoter −35 recognition. Fine mapping is in progress in collaboration with Hayward and colleagues.

but becomes exposed in the $\alpha_2\beta$ complex. If this were not true, the contact site to β' would be formed between the two subunits.

Subunit β' is a basic protein with non-specific DNA binding activity. Mutant RNA polymerases containing mutant β' subunits show irregular profile of RNA chain elongation, presumably due to altered binding to template DNA during transcription. Mapping of functional sites on β', including the subunit-subunit contact site(s), is in progress using a set of *rpoC* deletion mutants.

2. *Subunit σ Contact Sites on the Core Subunits*

Subunit σ^{70} with a molecular weight of 70 kDa is a major σ subunit in exponentially growing cells and is involved in transcription of house-keeping genes. Core enzyme subunits are assembled in the absence of σ^{70} subunit, but the core enzyme thus reconstituted *in vitro* in the absence of σ subunit is virtually inactive in RNA synthesis. Assembled but inactive core enzyme ("premature core enzyme") differs from active matured core enzyme in various structural parameters as measured by sedimentation profile, circular dichroism, tritium exchange rate, and protease sensitivity (*13*). Maturation of the premature core enzyme is promoted by the addition of σ^{70} subunit (*14*). Thus, the σ subunit has dual functions, *i.e.*, promotion of the core enzyme maturation leading to formation of the holoenzyme, and of promoter recognition by the holoenzyme. It remains to be determined, however, whether other minor σ subunits also function to promote core enzyme maturation.

Among isolated core enzyme subunits, the β' subunit alone is able to form a complex with σ^{70} (*15*) (Fig. 1). The isolated β' subunit forms aggregates, but in the presence of σ^{70} subunit it becomes soluble by forming β-σ^{70} complexes. This indicates that a protein-protein contact site exists on both of these subunits. C-terminally truncated α subunits are able to interact with σ^{70} subunit and form pseudo-holoenzymes, if they retain the N-terminal region for core enzyme assembly (*9*, *10*). Neither α nor β alone is able to form a complex with σ^{70}, but the $\alpha_2\beta$ complex becomes competent in interaction with σ^{70} (*15*). A deletion in the C-terminal proximal region of the β subunit renders the core enzyme unable to interact with the σ^{70} subunit (*12*) (see Fig. 2). These observations altogether suggest that the σ^{70} subunit interacts with not only β' but also β subunit. The σ^{70} contact site on β, however, is exposed only when it is assembled at least up to the $\alpha_2\beta$ complex.

3. Core Enzyme Contact Site on the σ Subunits

Sequence analysis of the genes coding for the σ family proteins from diverse prokaryotes indicated that four regions (regions 1 to 4 starting from the N terminus) exist (Fig. 2), each carrying an amino acid sequence that is conserved among σ subunits (*4*). Regions 2 and 4 can be further subdivided into four (2.1 to 2.4) and two (4.1 and 4.2) subregions (*16*). Most promoters recognized by the RNA polymerase holoenzyme carrying σ^{70} consist of two defined hexanucleotide sequences, referred to as -10 and -35 to indicate their DNA distance from the transcription initiation site. Genetic analyses suggest that two distinct regions of σ^{70} are responsible for recognition of these sequences (*17*, *18*): region 2 is involved in recognition of promoter -10; and region 4 recognizes promoter -35. In collaboration with R.S. Hayward and colleagues, detailed mapping of the region involved in promoter recognition is being carried out by making a set of deletion mutants in each conserved region (A. Kumar *et al.*, in preparation). The region in σ involved in binding the core enzyme was first thought to be located within region 2.2, because the region 2.2 sequence is conserved between some core enzyme binding proteins such as NusA. Deletion analysis of the σ^{70} subunit, however, indicated instead that a region upstream from region 2.2 is involved in core binding (*16*). As described above, the target site on the core enzyme of σ binding has not yet been determined. Further analysis of mutants, as well as cross-linking experiments, is needed to get a detailed map of the contact sites between the σ subunit and the core enzyme subunits.

4. Multiple Species of the σ Subunit

The first step in modulation of RNA polymerase takes place with replacement of the σ subunit. Up to the present time, seven different species of σ subunit have been identified in *E. coli*. Most of the housekeeping genes expressed in exponentially growing *E. coli* cells are recognized by the holoenzyme containing the major σ subunit, σ^{70}. Other minor species of σ are induced under stress conditions or at non-growing stationary phase, and holoenzymes containing these minor σ subunits are responsible for transcription of stress-induced or stage-specific genes (for review, see refs. *3*, *4*). In collaboration with Hayward and colleagues, we are searching for other, still unidentified σ subunits. For this purpose, we synthesized antibodies against a synthetic peptide with the sequence highly conserved at the σ subunit 2.2 region (Fig. 2). These antibodies cross-reacted with more than 10 different species of *E. coli* proteins besides the known σ subunits (*19*). We have to date purified four immunologically cross-reacting proteins against the sigma peptide antibodies (SCRP's) and sequenced them at their N-terminal regions. To clone the genes encoding the SCRPs, probes were designed based on knowledge of their N-terminal amino acid sequences. Three of the four purified SCRPs were, however, identified from their N-terminal sequences as thioredoxin reductase, ribosomal protein S2 and alkylhydroperoxide reductase (R. Ueshima, N. Fujita, and A. Ishihama, submitted for publication).

Another set of probes for the cloning of genes encoding minor σ was prepared, each carrying one of the conserved nucleotide sequences between different σ subunits. Using these two sets of probes and the DNA filter of the Kohara library of *E. coli* genome, some distinct loci which may represent hitherto unmapped genes have been identified (*20*).

II. COMMUNICATION BETWEEN RNA POLYMERASE AND TRANSCRIPTION FACTORS

1. Transcription Factor Contact Site I on the α Subunit

Functions of RNA polymerase are also modulated by interaction with accessory transcription factors. Two groups of transcription factors have been identified in *E. coli* (reviewed in ref. *3*). Factors in one group form RNA polymerase complexes in the absence of DNA; these include at least ten protein factors and two nucleotide factors (*3*). RNA polymerase-transcription factor complexes can be isolated simply by glycerol gradient centrifugation, but for some factors only under restricted conditions (A. Ishihama, in preparation). The mode and affinity of molecular interaction is different between transcription factors and between holo- and core enzymes. Factors interacting with the holoenzyme are considered to be involved in the modulation of promoter recognition properties of the holoenzyme, while those interacting with the core enzyme more or less influence the step of RNA chain elongation.

Transcription regulatory proteins such as repressors and activators have high affinity to target signals on DNA. Several lines of evidence, however, indicate that factors in this group also interact with RNA polymerase when both RNA polymerase and transcription factors bind to the same DNA molecules. Thus, regardless of apparent difference in the order of molecular assembly, ternary complexes are formed between RNA polymerase, a transcription factor and DNA signals at the final stage of transcriptive complex formation.

During the functional mapping of RNA polymerase subunits, we found that the C-terminal proximal region of α subunit is not necessary for both the RNA polymerase assembly and its catalytic function. Mutant RNA polymerases containing C-terminal truncated α subunits are able to transcribe simple promoters, which do not require additional transcription factors for function. However, the C-terminal region of α is needed for molecular communication between RNA polymerase and transcription factors such as CRP (cAMP receptor protein) and OmpR (a regulatory protein for the *omp* regulon), but only when these factors bind to their target sites which are located upstream of the promoters (*21*, *22*). Lack of direct physical interaction between CRP and mutant RNA polymerases consisting of C-terminal truncated α has been observed by DNA footprinting (A. Kolb and H. Buc, personal communication) and transfer of CRP-linked fluorescence (R. Ebright, personal communication). This site has been designated as "contact site I" for the protein-protein communication between RNA polymerase and transcription factors (*22*) (Fig. 3A).

2. *Fine Mapping of Contact Site I*

To precisely locate the contact site for CRP, we made a library of the mutant *rpoA* genes, each carrying a single mutation on average within the C-terminal half. Using this mutant collection, we isolated *E. coli* mutants which showed Lac$^-$ phenotype in the presence of CRP and cAMP. Sequence analysis of some of these mutations indicated that the *rpoA* mutations were clustered only within a small region of α (Fig. 3B), presumably at the contact site for CRP (*23*).

All the *rpoA* mutants except the two temperature-sensitive mutants are defective in the expression of a set of genes which, though not essential for normal cell growth, is required for survival under stress conditions. Some of these mutations have recently been sequenced, as summarized in Fig. 3B. For instance, the *rpoA*109 mutation at amino acid residue 290 interferes only with expression of late genes of phage P2 and its satellite P4, because molecular communication is blocked between the mutant RNA polymerase and positive activators, Ogr protein (P2) and δ protein (P4), for transcription of phage late genes (*24*, *25*). The *rpoA*341 (phs) mutation interferes with molecular communication between the mutant RNA polymerase and positive regulators (CysB, MelR, and AraC) required for expression of the *cysA*, *melAB*, and *araBAD* genes (*26*). This mutation has been found to be Lys to Phe change at amino acid residue 271 (*27*). The mutations *rpoA*77 and *rpoA*63 have been isolated independently in two groups (*28*, *29*). Both affect transcription of the

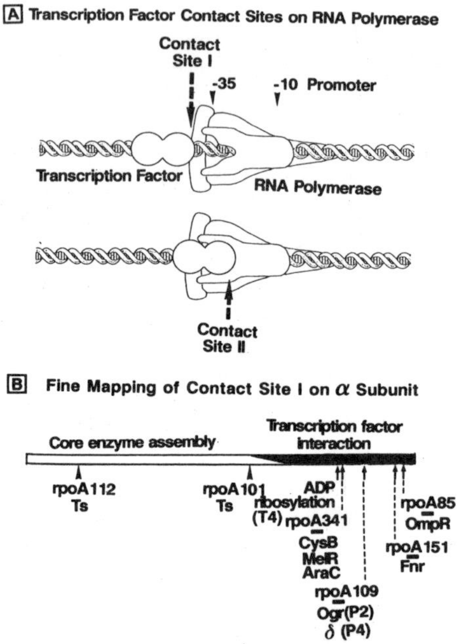

Fig. 3. Mapping of the contact sites on RNA polymerase with transcription factors. (A) Transcription factors that bind target signals upstream from the promoter interact with contact site I on the C-terminal region of α subunit. Factors that bind near the promoter region interact with contact site II. Location of site II has not yet been determined. (B) A number of mutants carrying point mutations in the *rpoA* gene were isolated in this laboratory which showed Lac^- phenotype and reduced levels of β-galactosidase production in the presence of cAMP and CRP. These mutations were all mapped within a small region of contact site I. Other *rpoA* mutants carrying point mutations within contact site I have been isolated and sequenced in other laboratories. All these mutants interfere with transcription of only one set of genes which require transcription factors as indicated for each mutation.

ompC and *ompF* porin protein genes which require an activator protein OmpR. The *rpoA*63 mutation and other *rpoA* alleles newly isolated by similar screening have been mapped in the C-terminal proximal region (*30*). All the *rpoA* mutations recently isolated by screening mutants defective in transcription of anaerobically regulated genes have also been mapped in the extreme terminus of the α subunit (*31*). These genes require a positive regulatory protein, Fnr, which is similar to CRP in the amino acid sequence. After phage T4 infection, *E. coli* RNA polymerase α subunit is ADP-ribosylated by phage T4-coded enzymes at the position of Arg^{265} within the contact site I located in the C-terminal proximal region (*32*). Taken together, these results demonstrate that multiple contact domains are clustered within contact site I of the

C-terminal one-third of the α subunit (Fig. 3), each being involved in communication with one of the site I-binding transcription factors. I would like to propose that these subsites within the contact site I be called "site I (CRP)", "site I (Fnr)" and "site I (OmpR)".

A battery of monoclonal antibodies against α subunit was produced by Krakow and colleagues (*33*). Some antibodies which recognize epitopes on the C-terminal proximal region competed with CRP for RNA polymerase binding (*34*; K. Sharif *et al.*, in preparation). This collection may be useful for fine mapping of subsites within contact site I for various transcription factors.

3. *Transcription Factor Contact Site II*

In contrast, transcription factors including CRP and PhoB, which bind at their target signals that overlap with promoter, were found to activate those mutant RNA polymerases containing C-terminal truncated α (*21*), suggesting that there is at least one other contact site for RNA polymerase-transcription factor communication. Mapping of the putative contact site II (and others) is in progress by analysis of suppressor mutations of activation-defective transcription factors.

SUMMARY

The RNA polymerase core enzyme with the subunit structure of $\alpha_2\beta\beta'$ constitutes the core machinery of transcription apparatus of *E. coli.* Its promoter selectivity is determined based on the molecular species of σ subunit. The promoter selectivity is, however, subject to modulation through molecular interaction with various accessory transcription factors.

The ordered assembly of RNA polymerase subunits is required for proper molecular contact between subunits and thus for expression of the catalytic activity of RNA polymerization. The first-step specificity determination of RNA polymerase takes place by binding of one of at least seven different species of σ subunits in *E. coli.* The existence of additional minor σ factors has been determined by immunological crossreaction of whole cell lysates with specific antibodies or by cross-hybridization with σ-specific probes.

The holoenzyme is further differentiated by interaction with various transcription factors. Class-I factors which mostly bind upstream of promoters make contact with the contact site I on RNA polymerase located at the C-terminal region of α subunit. Each contact site seems to be composed of a structure formed by a short segment only 5–10 amino acid residues in length. On the other hand, class-II factors, of which the binding sites on DNA overlap with promoters, interact with RNA polymerase at a site other than the α-subunit C-terminus.

Acknowledgments

The studies described in this report were carried out by K. Igarashi, R.S. Hayward, K. Sharif, C. Zou, and N. Fujita, and in collaboration with A. Hanamura and H. Aiba (Nagoya Univ.), K. Makino and A. Nakata (Osaka Univ.), H. Aiba and T. Mizuno (Nagoya Univ.), A. Kolb and H. Buc (Institut Pasteur), J.S. Krakow (Hunter College), M. Thomas and R.E. Glass (Univ. Nottingham), and R. Ebright (Rutgers Univ.). Research in Mishima was supported by Grants-in-Aid from the Ministry of Education, Science and Culture of Japan.

REFERENCES

1. T. Yura and A. Ishihama: *Annu. Rev. Genet.* **13**, 59–97 (1979).
2. A. Ishihama: *Adv. Biophys.* **21**, 163–173 (1986).
3. A. Ishihama: *Trends Genet.* **4**, 282–286 (1988).
4. J.D. Hellman and M.J. Chamberlin: *Annu. Rev. Biochem.* **57**, 839–872 (1988).
5. A. Ishihama: *Adv. Biophys.* **14**, 1–35 (1981).
6. A. Ishihama, N. Shimamoto, H. Aiba, K. Kawakami, H. Nashimoto, A. Tsugawa, and H. Uchida: *J. Mol. Biol.* **137**, 137–150 (1980).
7. K. Igarashi, N. Fujita, and A. Ishihama: *Nucl. Acids Res.* **18**, 5945–5948 (1990).
8. K. Kawakami and A. Ishihama: *Biochemistry* **19**, 3491–3495 (1980).
9. K. Igarashi, N. Fujita, and A. Ishihama: *J. Mol. Biol.* **218**, 1–6 (1991).
10. R.S. Hayward, K. Igarashi, and A. Ishihama: *J. Mol. Biol.* **221**, 23–29 (1991).
11. R.E. Glass, V. Nene, and M.G. Hunter: *Biochem. J.* **203**, 1–13 (1982).
12. R.E. Glass, A. Honda, and A. Ishihama: *Mol. Gen. Genet.* **203**, 492–495 (1986).
13. A. Ishihama, H. Aiba, T. Saitoh, and S. Takahashi: *Biochemistry* **18**, 979–983 (1979).
14. A. Ishihama, R. Fukuda, and K. Ito: *J. Mol. Biol.* **79**, 127–136 (1973).
15. R. Fukuda and A. Ishihama: *J. Mol. Biol.* **87**, 523–540 (1974).
16. S.A. Lesley and R.R. Burgess: *Biochemistry* **28**, 7728–7734 (1989).
17. T. Gardella, H. Moyle, and M.M. Susskind: *J. Mol. Biol.* **206**, 579–590 (1989).
18. D.A. Siegele, J.C. Hu, W.A. Walter, and C.A. Gross: *J. Mol. Biol.* **206**, 591–603 (1989).
19. N. Fujita, A. Ishihama, Y. Nagasawa, and S. Ueda: *Mol. Gen. Genet.* **203**, 492–495 (1986).
20. R. Ueshima, N. Fujita, and A. Ishihama: *Biochem. Biophys. Res. Commun.* **184**, 634–639 (1992).
21. K. Igarashi and A. Ishihama: *Cell* **65**, 1015–1022 (1991).
22. K. Igarashi, A. Hanamura, K. Makino, H. Aiba, H. Aiba, A. Nakata, and A. Ishihama: *Proc. Natl. Acad. Sci. U.S.A.*, **88**, 8958–8962 (1991).

23. C. Zou, N. Fujita, and A. Ishihama: *Mol. Microbiol.*, in press.
24. M.G. Sunshine and B. Sauer: *Proc. Natl. Acad. Sci. U.S.A.* **72**, 2770–2774 (1975).
25. C. Halling, M.G. Sunshine, K.B. Lane, E.W. Six, and R. Calendar: *J. Bacteriol.* **172**, 3541–3548 (1990).
26. P.M. Giffard and I.R. Booth: *Mol. Gen. Genet.* **214**, 148–152 (1988).
27. M. Thomas and R.E. Glass: *Mol. Microbiol.* **5**, 2719–2725 (1991).
28. S.-I. Matsuyama and S. Mizushima: *J. Mol. Biol.* **195**, 847–853 (1987).
29. S. Garrett and T.J. Silhavy: *J. Bacteriol.* **169**, 1379–1385 (1987).
30. J.M. Slauch, F.D. Russo, and T.J. Silhavy: *J. Bacteriol.* **173**, 7501–7510 (1991).
31. M.-L. Lombardo, D. Bagga, and C.G. Miller: *J. Bacteriol.* **173**, 7511–7518 (1991).
32. D. Rabussy: In "Bacteriophage T4" (Ed. by C.K. Mathews, E.M. Kutter, G. Mosig, and P.B. Berget), pp. 167–173, Am. Soc. Microbiol., Washington DC (1983).
33. F. Riftina, E. DeFalco, and J.S. Krakow: *Biochemistry* **28**, 3299–3305 (1989).
34. F. Riftina, E. DeFalco, and J.S. Krakow: *Biochemistry* **29**, 4440–4446 (1990).

Molecular Structure and Life (Ed. by Y. Kyogoku and Y. Nishimura), pp. 207–216, Japan Sci. Soc. Press, Tokyo/CRC Press, Boca Raton (1992)

Altered DNA-binding Specificity Mutants of GCN4 and TFIID

Kevin STRUHL

*Department of Biological Chemistry and Molecular Pharmacology, Harvard Medical School**

Mutant DNA-binding proteins with altered sequence recognition properties have been extremely useful for defining specific protein-DNA contacts mediated by the helix-turn-helix structural motif (*1–8*). Previously characterized specificity mutants contain single amino acid substitutions and alter DNA sequence recognition at a particular base pair, thus providing strong functional evidence for direct contacts between individual amino acids and base pairs. In general, direct interactions inferred from genetic and biochemical studies have been confirmed by high resolution structures of the protein-DNA complexes (*9–11*). This paper will review our studies (*12*, *13*) concerned with the isolation and characterization of altered specificity mutants of two transcriptional regulatory proteins from the yeast *Saccharomyces cerevisiae*, GCN4 and TFIID.

I. GCN4

GCN4 protein binds to the promoters of many amino acid biosynthetic genes and coordinately activates their transcription (*14*, *15*). Optimal

* Boston, MA 02115, U.S.A.

binding is observed with a 9-bp dyad symmetric sequence, ATGA(C/G) TCAT, with the central 7-bp being most important (*16*, *17*). GCN4 binds as a dimer to overlapping and non-equivalent half-sites (*18*), and the optimal half-site is ATGAC (*19*). The 56 C-terminal amino acids of GCN4 are sufficient both for dimerization and for specific DNA-binding (*18*, *20*, *21*), and this DNA-binding domain folds independently of the remainder of the protein (*22*).

The GCN4 DNA-binding domain contains the bZIP structural motif found in a class of eukaryotic transcription factors that includes C/EBP and the Jun and Fos oncoproteins (*23*). The bZIP domain is largely α-helical (*21*, *24–26*) and consists of a dimerization element, the leucine zipper, and an adjacent basic region that directly contacts DNA (*24*, *27–31*). As predicted by models of the protein-DNA complex (*25*, *32*), the leucine zipper symmetrically positions a diverging pair of α-helical basic regions to make sequence-specific contacts with the DNA target (*33*). The basic region contains a quartet of uncharged residues including two alanines, a serine/cysteine, and an invariant asparagine (corresponding to asn235, ala238, ala239, and ser242 in GCN4) that have been proposed to lie on the face of the α-helix that contacts the DNA (*25*). In support of this idea, DNA-binding activity is retained in a derivative in which 5 non-conserved residues on the putative solvent-exposed surface are changed to alanine or glutamine (*25*). Surprisingly, however, these highly conserved residues are not essential for DNA-binding by GCN4 (*34*).

As a general approach for identifying GCN4 derivatives with altered DNA-binding specificity, we utilized a set of yeast strains that differ only in the GCN4 recognition sequence upstream of the TATA element in the *his3* promoter. GCN4 binding to a functional target sequence activates *his3* transcription, which permits cells to grow in the presence of aminotriazole (AT), a competitive inhibitor of the *his3* gene product. Symmetrical double mutants representing equivalent changes in each of the adjacent half-sites (TTGACTCAA, ACGACTCGT, ATTACTAAT, ATGTCACAT) bind GCN4 poorly and are unable to support GCN4-dependent activation in yeast cells (*19*). To isolate potential altered specificity mutants, collections of GCN4 proteins, generated by mutagenesis *in vitro* with degenerate oligonucleotides, are screened for their ability to activate *his3* transcription from the symmetrically mutated target sequences.

The first example of an altered specificity mutant of GCN4 came from an analysis of the invariant asparagine (asn235). Although most substitutions of asn235 abolish DNA-binding, the trp235 protein displays nearly wild-type function, and the gln235 and ala235 proteins show detectable activity. The ability of an amino acid to functionally substitute for asn235 does not correlate with its preference for assuming the N-cap position of an α helix (*35*), thus arguing against a prediction (*32*) that the invariant asparagine forms an N-cap structure that permits the α-helical basic region to bend sharply around the DNA. The most interesting result was that the trp235 protein activated transcription from TTGACT-CAA, a sequence refractory to activation by wild-type GCN4. In contrast, when the target sequence was GTGACTCAC, the site that naturally occurs in the *his4* promoter, the asn235 protein activates strongly, whereas the trp235 protein appears inactive. Both proteins activate transcription from a promoter containing the optimal binding site, ATGACTCAT, although wild-type GCN4 is more efficient because it permits strains to grow at higher concentrations of aminotriazole.

DNA-binding experiments *in vitro* indicate that the asn235 protein binds more efficiently to the optimal (ATGACTCAT) site than the trp235 protein. GCN4 binds the native *his4* site (GTGACTCAC) almost as strongly as the optimal site, while the trp235 protein does not bind detectably. In contrast, the two proteins bind TTGACTCAA with comparable affinity. In terms of relative affinities, the order for GCN4, from highest to lowest binding, is ATGACTCAT > GTGACTCAC > ATGAC-TCTT > TTGACTCAA. For the trp235 protein, the order is TTGACT-CAA = ATGACTCAT > ATGACTCTT > GTGACTCAC. Analysis of the gln235 and ala235 proteins for their sequence preferences at the $\pm$4 position surprisingly reveal that both proteins favor the optimal site ATGACTCAT, bind weakly to TTGACTCAA and CTGACTCAG and strongly discriminate against GTGACTCAC. As expected from the *in vivo* results, the gln235 and ala235 proteins bind with lower affinity than the trp235 protein.

The trp235, gln235, and ala235 proteins can be interpreted as having lost some of the high affinity interactions at position $\pm$4 because all of these strongly discriminate against GTGACTCAC, a site efficiently bound by GCN4. This strongly suggests that asparagine 235 contributes to high-affinity GCN4 binding by recognizing, directly or indirectly, the $\pm$4 position. However, asn235 is clearly not responsible for all specificity at position $\pm$4, because the trp235, gln235, and ala235 substitutions retain some of the normal sequence preferences. On the other hand, the trp235 protein is not simply a "loss of specificity" mutant because it has the novel property of binding with comparable affinity to TTGACTCAA and the optimal site. Moreover, as tryptophan is the only substitution at position 235 that permits activation from TTGACTCAA, the altered specificity is not easily explained simply by the absence of asn235. Although indirect, conformationally transmitted effects cannot be excluded, we favor the idea of a direct contact between amino acid 235 and position ±4, especially because tryptophan, glutamine, and alanine are structurally very different yet all result in a strong discrimination against GTGACTCAC. Moreover, in a chimeric protein containing the Jun basic region in place of the GCN4 basic region, the only position 235 derivative that appears to activate transcription from TTGACTCAA is the equivalent change from asparagine to tryptophan.

Other protein-DNA contacts involved in high affinity binding have come from recently identified derivatives that display altered DNA-binding specificity at different positions of the binding site (J. Kim, D. Tzamarias, and K. Struhl, unpublished). At the +3 position, a protein containing tyr238 instead of the conserved ala238 has the novel property of activating transcription from ACGACTCGT. DNA-binding experiments indicate that the tyr238 protein can recognize this sequence, whereas the wild-type protein cannot. The tyr238 protein does not bind to any symmetrical double mutants at the ±2 or ±4 position indicating specificity at ±3. Interestingly, however, this protein retains the ability to bind the optimal GCN4 site. At the ±2 position, changing the moderately conserved ser242 to leu242 results in a protein that has near wild-type affinity for ATTACTAAT, a site not recognized by GCN4. Again, the ser242 protein is unable to recognize any symmetrical double mutants at ±3 or ±4, but retains the ability to bind the optimal GCN4 target.

Taken together, these observations strongly support a model in which the α-helical surface defined by amino acid positions 235, 238, and 242 is aligned along the DNA with direct contacts to ±4, ±3, and ±2, respectively. However, we do not understand the structural basis for why all of our altered specificity mutants retain the ability to recognize the optimal binding site. Hopefully, conclusive proof of the above hypothesis as well as more detailed knowledge of the chemical nature of the protein-DNA interactions will be addressed by a forthcoming high resolution structure of the protein-DNA complex (Tom Ellenberger and Steve Harrison, personal communication).

II. TFIID

TFIID is the highly conserved component of the RNA polymerase II transcription machinery that binds specifically to the TATA element (consensus TATAAA). The C-terminal 180 residues of the various

TFIIDs are at least 80% identical in amino acid sequence, and this core domain is necessary and sufficient for TATA-element binding and basal transcription *in vitro* (*36–39*), and for the essential functions of TFIID in yeast cells (*40–43*).

TFIID appears to be an unusual sequence-specific DNA-binding protein. It is extremely slow at binding to and dissociating from TATA elements (*44*), and it has a surprisingly high affinity for single-stranded DNA (*45*). Unlike most specific DNA-binding proteins, TFIID binds as a monomer (*37*) and it undergoes a conformational change upon binding to the TATA element (*39*). Optimal binding is not constrained to a simple target DNA site that conforms to the consensus of naturally occurring TATA elements, and a variety of non-consensus sequences interact efficiently with TFIID (*45*, *46*). Deletion analysis (*37*) and proteolytic cleavage experiments (*39*) indicate that the DNA-binding domain is not localized to a short region but instead requires the structural integrity of the entire core domain. Another unusual feature of the TFIID conserved core is the presence of two direct 67 amino acid repeats that are separated by a highly basic region. Dominant negative mutations in either repeat of yeast TFIID have been isolated that eliminate DNA binding while maintaining at least some aspects of normal structure and function (*42*). These mutations suggest that the TFIID monomer contains a bipartite DNA-binding domain in which each repeat contributes to sequence recognition, but they do not address the issue of binding specificity.

We have developed a new approach for addressing aspects of TFIID function that involves the isolation of derivatives that are transcriptionally active on mutated TATA elements (*13*). Such altered specificity mutants, which have been obtained for prokaryotic σ factors (*47–49*) provide strong genetic evidence for a direct protein-DNA contact. Moreover, by "genetically marking" TFIID with the property of altered TATA-element specificity, the activities of heterologous or mutated TFIIDs can be specifically assayed *in vivo* even in the presence of wild-type TFIID, which might be required for viability of the organism. Saturation mutagenesis of the yeast *his3* T_R TATA element revealed that almost all single base pair substitutions in the core sequence (TATAAA) severely compromised promoter function *in vivo* and *in vitro* (*50*, *51*). Consequently, cells carrying *his3* alleles with such defective TATA elements grow poorly on medium lacking histidine and not at all in the presence of aminotriazole (AT), a competitive inhibitor of the *his3* gene product. Thus, TFIID mutants with altered specificity for TATA elements would be isolated by virtue of their ability to increase transcription from a defective *his3* TATA element and hence to confer AT resistance.

To search for altered specificity mutants, the region of TFIID between amino acids 190 and 205 was heavily mutagenized by replacing it with a degenerate oligonucleotide containing 8% non-wild-type residues per base pair. This small region was targeted because it displays a weak sequence similarity (*52*) with a region of prokaryotic σ factors thought to interact with DNA (*53*). A library (10^6 independent clones) containing the collection of TFIID mutant proteins was introduced into a set of yeast strains with defective *his3* TATA elements (TGTAAA, GATAAA, TAGAAA, and TATAGA) whose transcriptional activities *in vitro* are at least 10-fold reduced compared to the wild-type TATAAA (*51*). All of these strains contain the wild-type TFIID gene on the chromosome to carry out the essential functions for cell growth.

From this genetic selection, four TFIID derivatives that permitted growth of the yeast strain containing the TGTAAA element were obtained; one of these, mutant 3, grew at higher drug concentrations suggestive of higher transcriptional activity. Indeed, a yeast strain carrying the TGTAAA element shows substantially increased *his3* RNA level in the presence of TFIID mutant 3 only. Moreover, this basal transcription is stimulated by GCN4, with the fold-induction being comparable to that expected for a wild-type promoter. The absolute level of *his3* transcription mediated by TFIID mutant 3 on the TGTAAA promoter is about 20% of that mediated by wild-type TFIID on an equivalent promoter containing TATAAA (*54*). Thus, the altered TFIID protein not only allows constitutive transcription to occur from a defective TATA element, but is also responsive to an acidic activator protein. Most importantly, the mutated TFIID derivatives do not function at defective TATA elements (GATAAA, TGTAAA, TCTAAA, TAGAAA, TATAGA) other than the TGTAAA sequence used in the original screening. Thus, as expected for an altered specificity mutant, these TFIID derivatives suppress mutations of the *his3* TATA element in an allele specific manner.

By analogy with altered specificity mutants of several DNA-binding proteins, TFIID derivatives functioning at mutated TATA sequences might be expected to be defective in activating transcription from promoters with natural TATA elements. Because the wild-type TFIID gene was present in these cells, the ability of the different mutants to carry out this and other essential functions of wild-type TFIID could not be tested. However, as determined by the plasmid shuffle complementation assay

(*55*), yeast cells containing mutants 1, 2, or 4 as the sole source of TATA-binding protein are viable, but they grow significantly less well than wild-type. For yeast cells containing mutant 3, very slow growing colonies are observed only after prolonged incubation. Thus, TFIID proteins bearing mutations that allow them to induce transcription from a defective TATA element are less efficient in promoting transcription from wild-type promoters. Moreover, the derivative allowing the highest *his3* expression from altered TATA element is also the least effective in carrying out the essential functions.

Surprisingly, all four TFIID altered specificity derivatives bear the same double amino acid substitution, ile$_{194}$ to phe$_{194}$ and leu$_{205}$ to val$_{205}$. Mutant 3, which supports the highest *his3* expression from the TGTAAA containing promoter, also contains a third amino acid change, val$_{203}$ to thr$_{203}$, that is not present in the other molecules. Substitutions of either glutamine (molecule 2) or asparagine (molecule 4) for the lysine at position 199 appear to be neutral because these molecules are phenotypically indistinguishable from molecule 1, which contains the wild-type residue. Given that all four altered specificity mutants share two particular base pair substitutions, it is very likely that any other pair of nucleotide changes within the mutagenized region could not yield TFIID derivatives that would pass the genetic selection employed here. Thus, increased activity from TGTAAA promoters cannot be accomplished with at least 80% of the possible single amino acid substitutions. The presence of the same double mutation in four independent mutants strongly suggests that both changes are required to produce a TFIID protein with altered specificity. Indeed, all other possible combinations of the mutations found in molecule 3 were completely unable to induce transcription from the TGTAAA containing promoter.

The DNA-binding specificities of wild-type TFIID and mutant 3 were directly examined on the complete set of TATA sequences that differ at position 2. As expected from *in vivo* (*50*) and *in vitro* (*51*) transcriptional analyses of the identical TATA sequences, wild-type TFIID binds efficiently to TATAAA, weakly to TTTAAA, and extremely poorly to both TGTAAA and TCTAAA. In striking contrast, the mutant protein binds much more efficiently than wild-type TFIID to the TGTAAA sequence. Otherwise, TFIID mutant 3 behaves very similarly to the wild-type protein on the four TATA sequences tested. Consistent with the observation that the mutant 3 stimulates transcription *in vivo* from TGTAAA with less than full wild-type efficiency, the mutant protein binds TGTAAA with somewhat lower affinity than TATAAA. These biochemical experiments directly demonstrate that TFIID mutant 3 displays altered DNA-binding specificity in that it gains the ability to efficiently bind TGTAAA while retaining other recognition properties of wild-type TFIID.

By analogy with altered specificity mutants of a variety of DNA-binding proteins, our results provide strong genetic evidence that the region of TFIID between residues 190 and 205 directly interacts with the TATA element. This very region had previously been implicated as being important for DNA binding because single substitutions at positions 196, 203, and 207 abolish TATA-element interaction while retaining at least some aspects of normal TFIID structure and function (*42*). In addition, a different substitution at position 205 is observed in the *spt15*-122 allele of TFIID, which alters the transcription pattern at the *his4*-917δ locus possibly by affecting TATA element utilization (*56*) (F. Winston, personal communication). Given that mutant 3 specifically increases binding to TGTAAA, the most likely explanation for altered specificity is the existence of a new contact between protein and position 2 of the TATA element that does not interfere with the normal TATA element interactions mediated by TFIID. However, the standard interpretation cannot be easily applied because altered TATA-element specificity requires two amino acid substitutions spaced 11 residues apart and is increased by a third substitution within the region. It is possible that residues 194, 203, and 205 are in close proximity in the folded structure of TFIID, with one or more of them directly interacting with position 2 of the TATA element. Alternatively, the mutations might subtly disrupt the conformation of a critical surface of TFIID and affect specificity more indirectly through another amino acid residue.

SUMMARY

Mutant DNA-binding proteins with altered sequence recognition properties are useful for defining specific protein-DNA contacts. This paper describes altered specificity mutants of two transcriptional regulatory proteins from the yeast *S. cerevisiae*, GCN4 and TFIID. TFIID is the highly conserved, but species-specific, component of the RNA polymerase II transcription machinery that binds specifically to the TATA element (consensus TATAAA). Using a genetic selection, we isolate an altered specificity derivative of yeast TFIID that permits transcription from promoters containing a mutated TATA element (TGTAAA). Biochemical analysis indicates that this TFIID derivative has specifically gained the ability to efficiently bind TGTAAA. The mutant protein contains three substitutions within a 12 amino acid region; two of these are necessary and primarily responsible for the altered specificity. These results define a surface of TFIID that directly interacts with the TATA element. By genetically selecting for derivatives of yeast GCN4 that activate transcription from promoters containing mutant binding sites, we isolate an altered-specificity mutant in which the invariant asparagine in the basic region of bZIP proteins (asn235) has been changed to tryptophan. Wild-type GCN4 binds the optimal site (ATGACTCAT) with much higher affinity than the mutant site (TTGACTCAA), whereas the trp235 protein binds these sites with similar affinity. Moreover, the trp235, ala235, and gln235 derivatives differ from GCN4 in their strong discrimination against GTGACTCAC. These results suggest a direct interaction between asn235 and the ± 4 position of the DNA target site. This and other specificity mutants will be very useful in interpreting a X-ray structure of the GCN4/AP-1 complex.

REFERENCES

1. P. Youderian, A. Vershon, S. Bouvier, R.T. Sauer, and M.M. Susskind: *Cell* **35**, 777–783 (1983).
2. R.H. Ebright, P. Cossart, B. Gicquel-Sanzey, and J. Beckwith: *Nature* **311**, 232–235 (1984).
3. A. Hochschild and M. Ptashne: *Cell* **44**, 925–933 (1986).
4. A. Hochschild, J.I. Douhan, and M. Ptashne: *Cell* **47**, 807–816 (1986).
5. R.P. Wharton and M. Ptashne: *Nature* **326**, 888–891 (1987).
6. N. Lehming, J. Sartorius, M. Niemoller, G. Genenger, B. vonWilcken-Bergmann, and B. Muller-Hill: *EMBO J.* **6**, 3145–3153 (1987).
7. S.D. Hanes and R. Brent: *Cell* **57**, 1275–1283 (1989).
8. J. Treisman, P. Gonczy, M. Vashishtha, E. Harris, and C. Desplan: *Cell* **59**, 553–562 (1989).
9. S.R. Jordan and C.O. Pabo: *Science* **242**, 893–899 (1988).
10. A.K. Aggarwal, D.W. Rodgers, M. Drottar, M. Ptashne, and S.C. Harrison: *Science* **242**, 899–907 (1988).
11. C.R. Kissinger, B. Liu, E. Martin-Blanco, T.B. Kornberg, and C.O. Pabo: *Cell* **63**, 579–590 (1990).
12. D. Tzamarias, W.T. Pu, and K. Struhl: *Proc. Natl. Acad. Sci. U.S.A.* **89**, in press.
13. M. Strubin and K. Struhl: *Cell* **68**, in press.
14. I.A. Hope and K. Struhl: *Cell* **43**, 177–188 (1985).
15. K. Arndt and G. Fink: *Proc. Natl. Acad. Sci. U.S.A.* **83**, 8516–8520 (1986).
16. D.E. Hill, I.A. Hope, J.P. Macke, and K. Struhl: *Science* **234**, 451–457 (1986).
17. A.R. Oliphant, C.J. Brandl, and K. Struhl: *Mol. Cell. Biol.* **9**, 2944–2949 (1989).
18. I.A. Hope and K. Struhl: *EMBO J.* **6**, 2781–2784 (1987).

19. J.W. Sellers, A.C. Vincent, and K. Struhl: *Mol. Cell. Biol.* **10**, 5077–5086 (1990).
20. I.A. Hope and K. Struhl: *Cell* **46**, 885–894 (1986).
21. M.A. Weiss, T. Ellenberger, C.R. Wobbe, J.P. Lee, S.C. Harrison, and K. Struhl: *Nature* **347**, 575–578 (1990).
22. I.A. Hope, S. Mahadevan, and K. Struhl: *Nature* **333**, 635–640 (1988).
23. W.H. Landschulz, P.F. Johnson, and S.L. McKnight: *Science* **240**, 1759–1764 (1988).
24. R.V. Talanian, C.J. McKnight, and P.S. Kim: *Science* **249**, 769–771 (1990).
25. K.T. O'Neil, R.H. Hoess, and W.F. DeGrado: *Science* **249**, 774–778 (1990).
26. L. Patel, C. Abate, and T. Curran: *Nature* **347**, 572–575 (1990).
27. W.H. Landschulz, P.F. Johnson, and S.L. McKnight: *Science* **243**, 1681–1688 (1989).
28. E.K. O'Shea, R. Rutkowski, and P.S. Kim: *Science* **243**, 538–542 (1989).
29. T. Kouzarides and E. Ziff: *Nature* **340**, 568–571 (1989).
30. J.W. Sellers and K. Struhl: *Nature* **341**, 74–76 (1989).
31. P. Agre, P.F. Johnson, and S.L. McKnight: *Science* **246**, 922–926 (1989).
32. C.R. Vinson, P.B. Sigler, and S.L. McKnight: *Science* **246**, 911–916 (1989).
33. W.T. Pu and K. Struhl: *Proc. Natl. Acad. Sci. U.S.A.* **88**, 6901–6905 (1991).
34. W.T. Pu and K. Struhl: *Mol. Cell. Biol.* **11**, 4918–4926 (1991).
35. J.S. Richardson and D.C. Richardson: *Science* **240**, 1648–1652 (1988).
36. T. Hoey, B.D. Dynlacht, M.G. Peterson, B.F. Pugh, and R. Tjian: *Cell* **61,** 1179–1186 (1990).
37. M. Horikoshi, T. Yamamoto, Y. Ohkuma, P.A. Weil, and R.G. Roeder: *Cell* **61**, 1171–1178. (1990)
38. M.G. Peterson, N. Tanese, F. Pugh, and R. Tjian: *Science* **248**, 1625–1630 (1990).
39. P.M. Lieberman, M.C. Schmidt, C.C. Kao, and A.J. Berk: *Mol. Cell. Biol.* **11**, 63–74 (1991).
40. B.P. Cormack, M. Strubin, A.S. Ponticelli, and K. Struhl: *Cell* **65**, 341–348 (1991).
41. G. Gill and R. Tjian: *Cell* **65**, 333–340 (1991).
42. P. Reddy and S. Hahn: *Cell* **65**, 349–357 (1991).
43. Q. Zhou, M.C. Schmidt, and A.J. Berk: *EMBO J.* **10**, 1843–1852 (1991).
44. M.C. Schmidt, Q. Zhou, and A.J. Berk: *Mol. Cell. Biol.* **9**, 3299–3307 (1989).
45. S. Hahn, S. Buratowski, P.A. Sharp, and L. Guarente: *Proc. Natl. Acad. Sci. U.S.A.* **86**, 5718–5722 (1989).
46. V.L. Singer, C.R. Wobbe, and K. Struhl: *Genes Dev.* **4**, 636–645 (1990).
47. T. Gardella, H. Moyle, and M.M. Susskind: *J. Mol. Biol.* **206**, 579–590 (1989).
48. D.A. Siegele, J.C. Hu, W.A. Walter, and C.A. Gross: *J. Mol. Biol.* **206**, 591–604 (1989).
49. P. Zuber, J. Healy, H.L.I. Carter, S. Cutting, C.P.J. Moran, and R. Losick: *J. Mol. Biol.* **206**, 605–614 (1989).
50. W. Chen and K. Struhl: *Proc. Natl. Acad. Sci. U.S.A.* **85**, 2691–2695 (1988).
51. C.R. Wobbe and K. Struhl: *Mol. Cell. Biol.* **10**, 3859–3867 (1990).
52. M. Horikoshi, C.K. Wang, H. Fujii, J.A. Cromlish, P.A. Weil, and R.G. Roeder: *Nature* **341**, 299–303 (1989).
53. J.D. Helmann and M.J. Chamberlin: *Annu. Rev. Biochem.* **57**, 839–872 (1988).
54. P.A.B. Harbury and K. Struhl: *Mol. Cell. Biol.* **9**, 5298–5304 (1989).
55. J.D. Boeke, J. Trueheart, G. Natsoulis, and G.R. Fink: *Methods Enzymol.* **154**, 164–175 (1987).
56. D.M. Eisenmann, C. Dollard, and F. Winston: *Cell* **58**, 1183–1191 (1989).

Molecular Structure and Life (Ed. by Y. Kyogoku and Y. Nishimura), pp. 217–227, Japan Sci. Soc. Press, Tokyo/CRC Press, Boca Raton (1992)

DNA Methylation and the Control of Gene Expression

Robin HOLLIDAY

*CSIRO Division of Biomolecular Engineering, Laboratory for Molecular Biology**

DNA contains more than four bases. In bacteria and their viruses, enzymic modification of bases occurs soon after DNA replication. Many types of modified bases are known, but the commonest are 6-methyl adenine and 5-methyl cytosine (5mC). In higher eukaryotes post-synthetic modification of cytosine to 5mC occurs primarily in CpG doublets. The DNA modification-restriction systems of bacteria demonstrate beyond doubt that enzymes can distinguish short sequences of DNA which contain methylated bases from those that do not. Restriction enzymes can cut DNA at these sequences if they are unmethylated, or if they contain a methylated base at an incorrect position. Modification enzymes methylate these sequences to protect the DNA from restriction. In some cases modification enzymes will methylate only the hemi-methylated DNA substrate, and not unmethylated DNA (*1*). This activity is known as maintenance DNA methylation, because it perpetuates through cell division a given pattern of methylation.

The mode of inheritance of DNA methylation and the fact that proteins could readily distinguish between methylated and unmethylated

* North Ryde NSW 2113, Australia

DNA was the primary basis for the suggestions 16 years ago that 5mC may have an important regulatory role in the control of gene expression in higher organisms (*2*, *3*). Since that time much information has been obtained about the significance of DNA methylation in many biological contexts, although there is no widespread agreement that it plays a primary or central role in unfolding the genetic programme for development. In this review I attempt to assess the significance of some of these observations, particularly those that relate to the concept of an epigenetic code consisting of 5 bases, superimposed on the genetic code of 4 bases.

I. DISTRIBUTION OF 5-METHYL CYTOSINE IN DNA

In higher organisms there is striking variation in the level of 5mC in different taxonomic groups. In insects the amount can be as low as 0.15–0.2% of all cytosine (*4*), and it is commonly asserted that *Drosophila* has none (see below). At the other extreme are many plant species where as much as 30% of cytosine in DNA is methylated, and in addition to CpG doublets, the sequence CpNpG also often contains 5mC (*5*). In vertebrates the proportion of modified cytosine is in the range 3–5%. The CpG doublet, which is very commonly methylated, is present at a much lower frequency than expected on a random basis. It has been suggested that the depletion of CpGs in the genome is due to the spontaneous deamination of 5mC to give thymidine, that is, a transition mutation converting CpG to TpG and its complement CpA. It is further argued that there is an equilibrium between the loss of CpG by deamination and the reappearance of CpG by mutation at other sites (*6*). This equilibrium model does not consider the possible functional significance of 5mC, and its preservation in DNA by selection. Recent analysis of human DNA demonstrates that the distribution of 5mC in specific tissues is highly conserved in the population (*7*, *8*). Although polymorphism in a methylation pattern has been documented (*9*), it appears to be much lower than predicted by the equilibrium model. It therefore seems very likely that the tissue specific pattern of methylation which is seen, and also the known changes of methylation in gametogenesis and embryogenesis (*10–12*), all have functional significance, although we do not know how the pattern of methylation is imposed and changed, nor in most cases its function.

There are regions of DNA which are not depleted in CpG doublets. These are known as CpG islands and they generally occur at the 5′ end of structural genes (*13*, *14*). The CpGs in islands are not methylated when the gene is expressed, but in the inactive X chromosomes islands have been shown to be methylated (reviewed in *15, 16*). Also, *de novo* methylation of island DNA in permanent cell lines leads to gene silencing (*17–20*). Plant DNA also contains CpG islands, although they are less obvious than those in mammalian DNA, and there is much less significant depletion of CpG or CpNpG sequences in plant DNA (*5, 21*).

II. HERITABILITY OF DNA METHYLATION

The concept of a maintenance methylase which strictly distinguishes between hemi-methylated DNA and non-methylated DNA remains, in part, a hypothesis. The major mammalian DNA transferase which has been purified and sequenced prefers a hemi-methylated substrate, but it does have *de novo* activity on non-methylated DNA (*22*). The evidence for the faithful transmission of the pattern of DNA methylation in cells comes from biological experiments. It is known that methylated DNA transfected into cells retains its methylation through many cell divisions, whereas the same sequences of non-methylated DNA remain non-methylated in cells. (Important exceptions to this are discussed below.) The inactive X chromosome is extremely stable through cell division and its CpG islands remain methylated, whereas the active X chromosome transmits non-methylated islands from cell to cell (*23*). This rule is, of course, broken at or just before meiosis, when the inactive X is normally reactivated (*15, 16*). In maize, transposable elements can exist in methylated or non-methylated form which are inherited in both somatic tissues and also through the germ line (*20, 24* and refs. therein). Many studies with permanent cell lines show that housekeeping genes can be in an active state associated with demethylated DNA and also in a silent or inactive form where CpG sites have become methylated, and these alternative states of genes are very stably inherited (*17, 20, 23, 25*). Finally, it is very well known that many latent viruses integrated into host DNA have methylated sequences which are perpetuated through cell division (reviewed in *17, 26*).

III. METHYLATION AND GENE EXPRESSION

There are many levels of genetic control, which are summarised in Table I. First, there is the basic DNA sequence which includes both the regulatory and coding regions of genes. Second, there are the non-sequence specific proteins which associate with DNA to form chromatin. Third, there are sequence specific transcription factors which interact

TABLE I
Levels of Genetic and Epigenetic Control

		Heritability
(1)	DNA sequence	+
(2)	Nucleosomes/chromatin	−
(3)	Sequence specific transcription factors	−
(4)	DNA methylation	+
(5)	Exons/introns, alternative splicing	(+)[a]

[a] The product of a spliced transcript may generate the same spliced combination of exons (*27*).

with regulatory regions of DNA, with each other and with RNA polymerase. Fourth, there is DNA methylation which can provide an inherited signal for one or more transcription factors. Fifth, there are additional mechanisms, such as alternative splicing, which can alter the primary structure of the gene product. Although suggestions have been made that particular forms of chromatin with or without associated transacting factors might be inherited as a protein-DNA complex, evidence that this is so has not been forthcoming (see ref. *20*). The significance of DNA methylation in the control of gene expression is that it provides a mechanism for the inheritance of a given pattern of gene activities in a somatic cell lineage or in a differentiated cell.

The first evidence that DNA methylation was related to gene activity was obtained using methyl sensitive restriction enzymes. If a DNA probe for the gene is available, it is possible to examine the methylated state of certain CpG residues in genomic DNA. This procedure has demonstrated in a large number of cases that methylation is associated with gene inactivity, and the converse, that active genes have demethylated DNA. The main limitation of the method is that only a subset of CpGs is examined, usually less than 10%, so the exact relationship of methylation to gene expression in a given region cannot be determined. This approach is now being superseded by genomic sequencing procedures which can distinguish between cytosine and 5mC at every position in a given sequence. The original method of Church and Gilbert (*28*) has been greatly improved and new methods have also been developed (*29–32*). Genomic sequencing has shown that the induction of the vitellogenin gene in chicken liver is associated with specific demethylation of two sites in the regulatory region (*33–35*). It has shown the specificity of the methylation pattern in particular tissues in different individuals (*7*, *8*) and it has also been successfully used to demonstrate the faithful inheritance of methylated and non-methylated CpG doublets in the CG island of the X linked phosphoglycerokinase (PGK) gene (*23*). These new experimental procedures will provide a wealth of new information about the relationship of DNA methylation to gene expression.

Another tool which has been widely used in the study of DNA methylation and gene activity is of more general application. The cytidine analogue 5-azacytidine (5-aza-CR) is incorporated into DNA and strongly inhibits the DNA methyl transferase activity (reviewed in *17*, *36*). This in turn can lead to extensive demethylation. Thus methylated genes which are silent can be reactivated at very high frequency by treatment with 5-aza-CR. 5-aza-CR can reactivate genes on the inactive X chromosome, silent genes in mammalian cell lines, or latent viruses (*17*). It can also induce the MyoD regulatory gene in 10T1/2 cells by the demethylation of a CpG island (*18*). It appears that the effect of 5-aza-CR is random, in that the pattern of demethylated sites produced varies from cell to cell, and that the "segmental" distribution of methylated and demethylated sites is subsequently inherited (*23*).

IV. IS DNA METHYLATION A PRIMARY SIGNAL?

In many publications it is stated that DNA methylation is correlated with lack of gene expression, or that it is a secondary consequence of some other regulatory event. There are many lines of evidence which suggest that the presence or absence of DNA methylation is a primary signal in gene regulation. There are now several examples of DNA sequence specific protein transcription factors which have very different binding affinities to the methylated or non-methylated substrate (*37–42* and refs. therein). Methylation of CpG within the target region may either promote or inhibit binding by the sequence specific protein.

In studies of maize transposable elements, it has been shown that certain inactive alleles have the normal DNA sequence, but contain methylated cytosine residues. These inactive states can be transmitted through meiosis to the following generation, and the expected phenotype is expressed (*24*). Thus, the inheritance of methylation is exactly analogous to the effect of a mutation (that is, a change in DNA base sequence) on the phenotype. It is not correct to state that the mutation is *correlated* with the phenotype because we know that the mutation causes the phenotype, and the same is true of the methylation differences.

There are several examples of specific DNA demethylation in specialised cell types, in which the gene is normally active (see Table II). These include the demethylation of DNA regulating the vitellogenin gene,

TABLE II
Specific Demethylation and Protein Synthesis

Protein	Cell type	References
δ Crystallin	Lens	(*43*)
Vitellogenin	Hen hepatocytes	(*33–35*)
α Actin	Myoblasts	(*44, 45*)
Ig κ light chain	B lymphocytes	(*46*)
Insulin	Insulinoma	(*46*)
Collagen IV	F9 teratocarcinoma	(*47*)

a crystallin gene, an actin gene, the insulin gene, an immunoglobulin gene and a collagen gene (*33–35, 43–47*). In some cases it has also been shown that the introduction of methylated DNA into an inappropriate cell, that is, one which does not express the gene, is not followed by demethylation. These experiments provide examples of a specific demethylation switch associated with a particular differentiated cell.

In the example of X chromosome inactivation, it has been proposed that DNA methylation provides a memory perpetuating the inactive state after an earlier inactivating event (*48*). This was supported by studies showing that the HPRT gene became methylated about 3 days later than the known time of inactivation. However, studies of the PGK gene show that in this case methylation occurs at the same time as X chromosome inactivation (*49*). There is no reason why inactivation should be instantaneous at all sites; indeed, there are quite strong grounds for believing that it is at least a two stage process, perhaps an initial switch followed by a spreading of inactivation along the chromosome over a period of a few cell divisions (*20*).

Another experimental approach has recently been used to obtain further evidence that DNA methylation might be a primary signal for gene silencing. 5-Methyl deoxycytidine triphosphate (5-methyl dCTP) was introduced into V79 or Chinese hamster ovary (CHO) cells by electroporation (*50, 51*). Normally cells have special mechanisms to prevent the incorporation of 5-methyl dCTP into DNA (*52*), but in the electroporation experiments the nucleotide is made available and appears to be directly incorporated into DNA. As a consequence, three housekeeping genes which were examined were all inactivated at a high frequency. These silenced genes can subsequently be reactivated by 5-aza-CR. The experiments strongly suggest that there is a direct relationship between DNA methylation and gene inactivity.

In studies of the Herpes thymidine kinase (TK) gene integrated into the DNA of human cells, clear evidence was obtained that a change in the methylation of only one or two cytosines in the promoter region was sufficient for a complete switch in gene expression (*53*, *54*). It should also be noted that in some of the earliest molecular studies of promoter regions of genes in mammalian cells, the same TK sequence was manipulated (*55*). Experiments of this type in which promoter sequences are mutated or deleted use DNA cloned in bacteria, which is then reintroduced into an appropriate mammalian cell to study transcription. Innumerable experiments of this type using many altered promoter sequences have subsequently been published. All these studies ignore the possible importance of DNA methylation in regulation, since all the promoter constructs tested are all demethylated.

V. DNA METHYLATION AND DEVELOPMENT

It is not widely believed that methylation has an important role in unfolding the developmental programme. One reason for this is the apparent absence of DNA methylation in *Drosophila*, the development of which is better understood than in any other organism. The best published data show that the level of methylation is less than 0.003% (*4*). A quantity of 0.001% corresponds to about 700 5mC residues in the genome, which is a sufficiently large number to provide regulatory signals. Also, there has been no thorough search for 6-methyl adenine or any other modified base in *Drosophila* DNA.

In mammals, there are several grounds for believing that methylation is important in development. It is directly implicated in X chromosome inactivation, and more recently in the fragile X syndrome (reviewed in *56*). Failure to reactivate a gene on the X chromosome may be one component of that syndrome (*57*). It is also associated with genomic imprinting, that is, the differences between male and female gametes which are not due to different DNA sequences (reviewed in *58*). The examples in Table II show that specialised or luxury genes are demethylated and activated in the appropriate cell context. Differentiation of F9 teratocarcinoma cells, including demethylation and transcription of the α1 (IV) collagen gene, is induced by retinoic acid, which is a powerful morphogenetic determinant.

Demethylation by 5-aza-CR can induce 10T1/2 cells to differentiate into muscle cells, adipocytes and chondrocytes (*36*). In the best studied muscle differentiation pathway, it is evident that the switch is induced by demethylation of a CG island of the MyoD gene. However, examination of the MyoD gene in various tissues showed that the island was in all cases unmethylated (*18*). These observations were based on the use of methyl sensitive restriction enzymes, which screen a minority of all CG doublets. It remains a possibility that the MyoD gene in muscle is unmethylated, and that in other tissues it has a subset of methylated CGs. This could now be examined using one of the new procedures of "5-base" genomic sequencing.

There are striking changes in the levels of methylation in DNA in gametogenesis and embryogenesis (*10–12*). Considering the importance of these stages of the life cycle, it would be very surprising if methylation did not have some functional role in genetic regulation or control. However, the significance of the changes seen is not understood. Another example of the possible importance of methylation in development comes from a study of vernalization in plants. Early flowering is induced in many species by cold treatment of seeds which are just beginning to germinate. It has been shown that 5-aza-CR treatment can effectively substitute for low temperature, suggesting that the demethylation of DNA is the important inducing signal (*59*). This signal would subsequently be inherited in the plant tissue and bring about early flowering.

CONCLUSIONS

Although much information about DNA methylation and gene expression in higher organisms has been published in the last 15 years or so, many enigmas remain. We do not know why there is so much 5mC in the DNA outside CG islands. It is very unlikely that all these 5mC residues have a regulatory role, since the CpG doublet itself is not a specific sequence. Nevertheless, it appears that the pattern of methylation is strictly controlled since it does not vary between individuals. Although the gene for the major methyl transferase has been cloned and sequenced, the biochemistry of the control of methylation in cells is obscure. It is possible that one domain of the methyl transferase interacts with other proteins to provide, on the one hand, accurate maintenance of pre-existing methylation, and on the other, specific *de novo* methylation at particular sequences. The interaction of the DNA methylase with sequence specific transcription factors is an obvious possibility. Another is that there may be more than one DNA transmethylase. A further problem is the mechanism of demethylation, which appears to occur in the absence of DNA synthesis. Advances are now being made in the study of proteins which have different interactions with specific methylated and non-methylated sequences (*37–42*), or which bind more generally to methylated DNA (*60–62*), but it should be noted that the innumerable published studies of transcription factors and the protein complexes they form almost always ignore the possibility that DNA methylation is an important signal in the control of gene transcription.

The original proposals that DNA methylation may have an important role in development were based in large part on the realisation that there must be epigenetic mechanisms which can on the one hand control the changes in the pattern of gene activity during development, and on the other provide a basis for the stable inheritance of the specialised phenotypes of dividing differentiated cells. Although much has been achieved, these basic epigenetic mechanisms are not yet understood.

SUMMARY

The properties of DNA can be strikingly altered by the methylation of bases in specific sequences, as is demonstrated by the restriction-modification systems of bacteria. In eukaryotes the commonest modified base is 5 mC. The pattern of modification varies between different tissues, or between homologous chromosomes (as in X chromosome inactivation or genomic imprinting), but it is largely invariant between different individuals of the same species. Much evidence shows that the methylation of DNA provides a heritable system for somatic cells, which is superimposed on the normal four base sequence in DNA. There is also evidence that the activation of specific genes in specialised cells is associated with the specific loss of methylation in regulatory regions of those genes.

It is stmetimes asserted that methylation is merely correlated with gene inactivity, or is a secondary consequence of other regulatory events. There are many published results that indicate that there is a direct relationship between the methylation of DNA and gene expression. Recent experiments show that the incorporation of 5-methyl dCTP into DNA leads to the silencing of genes in CHO and V79 cells. These results indicate that methylation can be a primary cause of gene inactivation.

The variety of biological contexts in which the alternative DNA bases cytosine or 5-mC have been related to changes in gene activity, strongly support the view that DNA methylation can provide a molecular basis for epigenetic controls of gene activity during the development of higher organisms.

REFERENCES

1. J.E. Kelleher, A.S. Daniel, and N.E. Murray: *J. Mol. Biol.* **221**, 431–440 (1991).
2. R. Holliday and J.E. Pugh: *Science* **187**, 226–232 (1975).
3. A.D. Riggs: *Cytogenet. Cell Genet.* **14**, 9–25 (1975).
4. C.V. Patel and K.P. Gopinathan: *Anal. Biochem.* **164**, 164–169 (1987).
5. M. Gardiner-Garden and M. Frommer: *J. Mol. Evol.* **34**, 231–245 (1992).
6. J. Sved and A. Bird: *Proc. Natl. Acad. Sci. U.S.A.* **87**, 4692–4696 (1990).
7. S. Konachek, M. Toth, A. Dehmel, D. Renz, and W. Doerfler: *Proc. Natl. Acad. Sci. U.S.A.* **87**, 8830–8834 (1990).
8. A. Behn-Krappa, I. Hoelker, U.S. de Silva, and W. Doerfler: *Genomics* **11**, 1–7 (1991).
9. A.J. Silva and R. White: *Cell* **54**, 145–152 (1988).
10. M. Monk, M. Boubelik, and S. Lehnart: *Development* **99**, 371–381 (1987).
11. J.R. Chaillet, T.F. Vogt, D.R. Beier, and P. Leder: *Cell* **66**, 77–83 (1991).
12. D. Frank, I. Keshet, M. Shani, A. Levine, A. Razin, and H. Cedar: *Nature* **351**, 239–241 (1991).
13. A.P. Bird: *Nature* **321**, 209–213 (1986).
14. M. Gardiner-Garden and M. Frommer: *J. Mol. Biol.* **196**, 261–282 (1987).
15. M. Monk: *BioEssays* **4**, 204–208 (1986).
16. S.G. Grant and V.M. Chapman: *Annu. Rev. Genet.* **22**, 199–233 (1988).
17. R. Holliday: *Science* **238**, 163–170 (1987).
18. P.A. Jones, M.J. Wolkowicz, M.A. Harrington, and F. Gonzales: *Phil. Trans. Roy. Soc. Lond. B* **326**, 277–284 (1990).
19. F. Antequera, J. Boyes, and A. Bird: *Cell* **62**, 503–514 (1990).
20. R. Holliday: *Biol. Rev.* **65**, 431–471 (1990).
21. M. Gardiner-Garden, J.A. Sved, and M. Frommer: *J. Mol. Evol.* **34**, 219–230 (1992).
22. T. Bestor, A. Landano, R. Matteliano, and V. Ingram: *J. Mol. Biol.* **202**, 971–983 (1988).
23. G.P. Pfeifer, S.D. Steigerwald, R.S. Hansen, S.M. Gartler, and A.D. Riggs: *Proc. Natl. Acad. Sci. U.S.A.* **87**, 8252–8256 (1990).
24. N. Federoff, P. Masson, and J.A. Banks: *BioEssays* **10**, 139–144 (1989).
25. R. Holliday and T. Ho: *New Biol.* **2**, 719–726 (1990).
26. W. Doerfler: *Phil. Trans. Roy. Soc. Lond. B* **326**, 253–265 (1990).
27. Z. Zachar, T.B. Chan, and P.M. Bingham: *EMBO J.* **6**, 4105–4111 (1987).
28. G. Church and W. Gilbert: *Proc. Natl. Acad. Sci. U.S.A.* **81**, 1991–1995 (1984).
29. H.P. Saluz and J.P. Jost: "A Laboratory Guide to Genomic Sequencing," Birkhausen, Boston (1987).
30. G.P. Pfeifer, S.D. Steigerwald, S. Mueller, P.R. Wold, and A.D. Riggs: *Science* **246**, 810–813 (1989).
31. G.P. Pfeifer, R. Taugnay, S.D. Steigerwald, and A.D. Riggs: *Genes & Dev.* **4**, 1277–1287 (1990).
32. M. Frommer, L.E. McDonald, D.S. Miller, C.M. Collis, F. Watt, G.W. Grigg, P. Molloy, and C.L. Paul: *Proc. Natl. Acad. Sci. U.S.A.*, in press.
33. H.P. Saluz, I.M. Feevers, J. Jiricny, and J.P. Jost: *Proc. Natl. Acad. Sci. U.S.A.* **85**, 6697–6700 (1988).

34. H.P. Saluz, J. Jiricny, and J.P. Jost: *Proc. Natl. Acad. Sci. U.S.A.* **83**, 7167-7171 (1986).
35. J.P. Jost, H.P. Saluz, I. McEwan, M. Hughes, S. Reiver, H.M. Liang, and M. Vaccaro: *Phil. Trans. Roy. Soc. Lond. B* **326**, 231-240 (1990).
36. P.A. Jones: *Cell* **40**, 485-486 (1985).
37. S.M.M. Iguchi-Ariga and W. Schaffner: *Genes & Dev.* **3**, 612-619 (1989).
38. R. Hermann, A. Hoeveler, and W. Doerfler: *J. Mol. Biol.* **210**, 411-415 (1989).
39. M. Comb and H.M. Goodman: *Nucl. Acids Res.* **18**, 3975-3982 (1990).
40. K. Ehrlich and M. Ehrlich: *Mol. Cell. Biol.* **10**, 4957-4960 (1990).
41. P.L. Molloy and F. Watt: *Phil. Trans. Roy. Soc. Lond. B* **326**, 267-276 (1990).
42. G.C. Prendergast and E.B. Ziff: *Science* **251**, 186-189 (1991).
43. C.H. Sullivan and R.M. Grainger: *Proc. Natl. Acad. Sci. U.S.A.* **84**, 329-333 (1987).
44. J. Yisraeli, R. Adelstein, D. Mellant, U. Nudel, D. Yoffe, and H. Cedar: *Cell* **46**, 409-416 (1986).
45. Z. Paroush, I. Keshet, J. Yisraeli, and H. Cedar: *Cell* **63**, 1229-1237 (1990).
46. D. Frank, M. Lichenstein, Z. Paroush, Y. Bergman, M. Shani, A. Razin, and H. Cedar: *Phil. Trans. Roy. Soc. Lond. B* **326**, 241-252 (1990).
47. P.D. Burbelo, S. Horikoshi, and Y. Yamada: *J. Biol. Chem.* **265**, 4839-4843 (1990).
48. L.F. Lock, N. Takagi, and G.R. Martin: *Cell* **48**, 39-46 (1987).
49. J. Singer Sam, M. Grant, J.M. Lebon, K. Okuyama, V. Chapman, M. Monk, and A.D. Riggs: *Mol. Cell. Biol.* **10**, 4987-4989 (1990).
50. J. Nyce: *Somatic Cell Mol. Genet.* **17**, 543-550 (1991).
51. R. Holliday and T. Ho: *Somatic Cell Mol. Genet.* **17**, 537-542 (1991).
52 J.A. Vilpo and L.M. Vilpo: *Mutat. Res.* **256**, 29-35 (1991).
53. J. Tasseron-de Jong, H. der Dulk, P. van de Putte, and M. Giphart Gassler: *Biochim. Biophys. Acta* **1007**, 215-223 (1989).
54. J. Tasseron-De Jong, J. Aker, H. der Dulk, P. van de Putte, and M. Giphart Gassler: *Biochim. Biophys. Acta* **1008**, 62-70 (1989).
55. S.L. McKnight: *Nucl. Acids Res.* **8**, 5949-5964 (1980).
56. T. Webb: *Curr. Biol.* **1**, 293-295 (1991).
57. C.D. Laird: *Genetics* **117**, 587-599 (1987).
58. D. Solter: *Annu. Rev. Genet.* **22**, 127-146 (1988).
59. W.J. Peacock: personal communication.
60. R.R. Meehan, J.D. Lewis, S. McKay, E.L. Kleiner, and A.P. Bird: *Cell* **58**, 499-507 (1987).
61. P. Supakau, D. Weist, D. Zhang, N. Inamdar, X.Y. Zhang, R. Khan, K.C. Ehrlich, and M. Ehrlich: *Nucl. Acids Res.* **16**, 8029-8044 (1988).
62. J.P. Jost, H.P. Saluz, and A. Pawlak: *Nucl. Acids Res.*, in press.

Molecular Structure and Life (Ed. by Y. Kyogoku and Y. Nishimura), pp. 229–235, Japan Sci. Soc. Press, Tokyo/CRC Press, Boca Raton (1992)

Supercoiling of DNA Facilitates TFIID-promoter Interactions

Susumu HIROSE, Hisahiro TABUCHI, and Mitsuko MIZUTANI

*DNA Research Center, National Institute of Genetics**

In prokaryotes, the superhelical state of DNA plays an important role in genetic processes including replication, transcription and recombination (see refs. *1* and *2* for reviews). However, little is known about such a role in eukaryotes. Though chromatin DNA in eukaryotic cells is topologically underwound, bulk DNA is not under superhelical tension because the supercoils are constrained by histone-DNA interactions (*3*). This does not necessarily exclude the possibility that unconstrained supercoils are present in a small fraction of the chromatin and contribute to execution of DNA functions. Consistent with this notion, an extract prepared from posterior silk gland cells of *Bombyx mori* contains an activity which introduces unconstrained negative supercoils into closed circular DNA (*4–6*). Purification of the activity reveals that it consists of DNA topoisomerase II and a supercoiling factor (*7*). Using this activity, we have shown that *in vitro* transcription of certain eukaryotic genes is activated by supercoiling of template DNA (*4–6*, *8*). This paper reviews the subsequent studies in this laboratory regarding the transcription enhancement by DNA supercoiling.

* Mishima, Shizuoka 411, Japan

I. TRANSCRIPTION OF DNA WITH DEFINED SUPERHELICITY

General transcription factors (TFIIB, TFIID, and TFIIE) and RNA polymerase II were partially purified from HeLa cells to reconstitute a transcription system. During the purification, care was taken to avoid contamination with DNA topoisomerase I and hence these transcription components were virtually free from topoisomerase I activity. Though the preparation of TFIIE contained DNA topoisomerase II activity, it was inactivated by a brief heat treatment without loss of TFIIE activity (*9*). When transcription reactions were carried out using TFIIB, TFIID, RNA polymerase II, and heated TFIIE, the linking number of closed circular DNA was maintained during transcription (*9*).

DNA templates containing defined negative superhelical densities were prepared by incubating plasmid DNA with DNA topoisomerase I in the presence of varying amounts of ethidium bromide. Each topoisomer population was tested for transcription activity in the reconstituted system described above and the results are shown in Fig. 1. Transcription of the *B. mori* fibroin gene on pFb205 DNA increases as the template superhelicity (σ value) changes from 0 to -0.036 and then retains the maximum level upon further increase in negative superhelicity of the template (Fig. 1A, the σ value of plasmid DNA isolated from *Escherichia coli* is approximately -0.06). Transcription of the adenovirus 2 major late promoter (Ad2 MLP) on pML (C2AT) DNA also increases as the σ value changes from 0 to -0.036 but then decreases with further increase in negative superhelicity (Fig. 1B), whereas transcription of the *Drosophila*

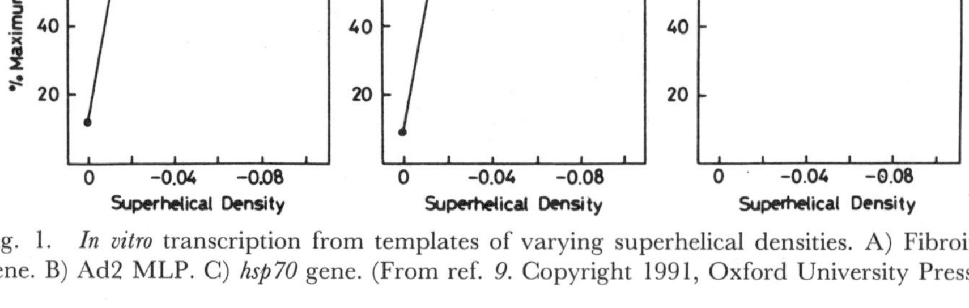

Fig. 1. *In vitro* transcription from templates of varying superhelical densities. A) Fibroin gene. B) Ad2 MLP. C) *hsp70* gene. (From ref. *9*. Copyright 1991, Oxford University Press)

hsp70 gene on p56H8RIA DNA appears to be unaffected by negative superhelicity of the template (Fig. 1C). These results illustrate the different effects of DNA superhelicity on the transcription of the three genes.

II. DNA SUPERHELICITY AFFECTS THE FORMATION OF TRANSCRIPTION PREINITIATION COMPLEX

To examine which step of transcription is affected by DNA supercoiling, we dissected transcription into pre- and post-initiation steps. First, we investigated the effect of DNA topology on the assembly of preinitiation complex using transcription components lacking DNA topoisomerase activities. DNA was preincubated with transcription factors and RNA polymerase II. At various times, 0.01% Sarkosyl was added to prevent further formation of the preinitiation complex. The extent of complex formation at each time was then measured by allowing a single round of transcription. As shown in Fig. 2A, superhelical DNA ($\sigma = -0.06$) supports more rapid formation of the preinitiation complex on the fibroin gene promoter than relaxed DNA does. Similar analysis of the Ad2 MLP revealed that formation of the preinitiation complex proceeds more rapidly on partially superhelical DNA ($\sigma = -0.036$) than on relaxed DNA (Fig. 2B). Note that the superhelical template of this σ value gives the maximum level of transcription from this promoter (see section above). The assembly of preinitiation complex on superhelical DNA with

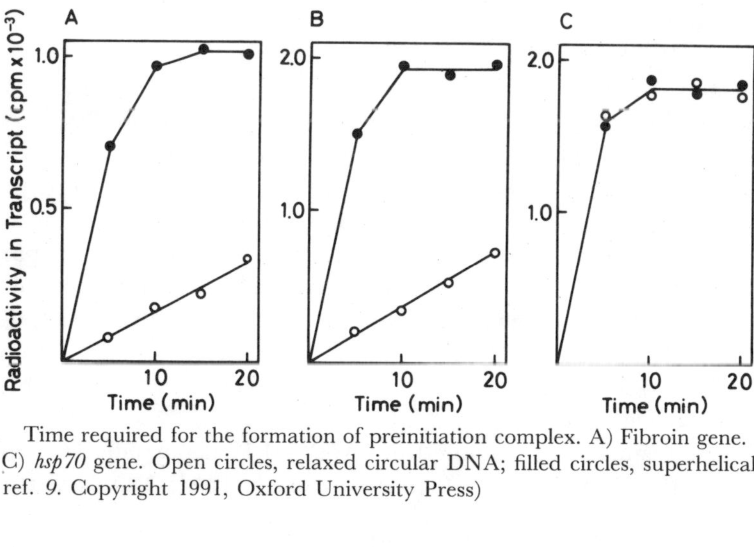

Fig. 2. Time required for the formation of preinitiation complex. A) Fibroin gene. B) Ad2 MLP. C) *hsp70* gene. Open circles, relaxed circular DNA; filled circles, superhelical DNA. (From ref. *9*. Copyright 1991, Oxford University Press)

σ value of -0.06 occurs faster than that on relaxed DNA but more slowly than that on partially superhelical DNA with σ value of -0.036 (data not shown). In the case of *hsp70* gene, the preinitiation complex assembles rapidly on both superhelical ($\sigma = -0.06$) and relaxed DNA (Fig. 2C).

When transcription was started synchronously from the preinitiation complex assembled on the fibroin gene promoter, the S1 nuclease protection band of expected size initially could not be seen but became detectable within 1 min after addition of nucleotides on both superhelical and relaxed DNA (data not shown). Similar results were obtained for transcription of the Ad2 MLP and the *hsp70* gene. Thus, the start of RNA chain polymerization and the subsequent elongation appear to be rapid processes irrespective of DNA topology once the preinitiation complex is formed. From these results we conclude that DNA superhelicity enhances transcription of certain genes by accelerating the formation of the transcription preinitiation complex, but does not affect transcription of other genes on which the preinitiation complex assembles rapidly even in the relaxed state.

III. DNA SUPERHELICITY FACILITATES INTERACTION OF TFIID TO THE FIBROIN GENE AND THE Ad2 MLP

To determine which transcription factor is responsible for the rapid assembly of preinitiation complex on the superhelical template each transcription factor was incubated with template DNA for 10 min. After addition of the remaining factors, the incubation was continued and 30 sec later 0.01% Sarkosyl was added. The amounts of preinitiation complex formed were measured by a single round of transcription. When superhelical DNA ($\sigma = -0.060$) carrying the fibroin gene promoter was preincubated with TFIID, the amounts of transcript were almost the same level as control where all transcription components were present during the preincubation (Table I, No. 3 *versus* No. 1). Similar results were obtained when purified yeast TFIID was used in place of partially purified human TFIID (data not shown). No significant level of transcript was detectable if the preincubation contained only TFIIB, TFIIE, or RNA polymerase II (No. 2, 4, or 5). Only a small amount of preinitiation complex was formed on relaxed DNA even if it had been preincubated with TFIID (No. 6).

Similar results were obtained for Ad2 MLP, if partially superhelical DNA of pML (C2AT) ($\sigma = -0.036$) was used (Table I, Nos. 7 through

TABLE I
Requirements for Rapid Assembly of Preinitiation Complex

No.	Protein added at 0 time	Template	UMP incorporated (fmol)
1	Complete	pFb205, sc	52
2	TFIIB	pFb205, sc	4
3	TFIID	pFb205, sc	46
4	TFIIE	pFb205, sc	3
5	RNA polymerase II	pFb205, sc	4
6	TFIID	pFb205, rc	9
7	Complete	pML(C2AT), psc	104
8	TFIIB	pML(C2AT), psc	7
9	TFIID	pML(C2AT), psc	92
10	TFIIE	pML(C2AT), psc	6
11	RNA polymerase II	pML(C2AT), psc	8
12	TFIID	pML(C2AT), rc	21

sc, supercoiled DNA ($\sigma=-0.06$); psc, partially supercoiled DNA ($\sigma=-0.036$); rc, relaxed circular DNA.

12). These results suggest that the functional binding of TFIID to the promoter is a rate limiting step in the assembly of preinitiation complex on the fibroin gene or the Ad2 MLP, and that DNA superhelicity facilitates the step.

IV. BINDING OF TFIID TO THE Ad2MLP INDUCES UNDERWINDING OF THE DNA HELIX

What is the mechanism by which negative supercoiling of DNA affects the TFIID-promoter interaction? The binding of TFIID to a promoter might induce underwinding of the DNA helix which is facilitated by DNA supercoiling. To test this possibility, partially superhelical DNA carrying the Ad2MLP ($\sigma=-0.036$) was incubated with yeast TFIID which had been expressed in and purified from *E. coli*. After addition of DNA topoisomerase I followed by further incubation, DNA was deproteinized and distribution of topoisomers was analyzed by two-dimensional electrophoresis. Control sample incubated without TFIID gave topoisomers with slight positive superhelical turns, because relaxation with DNA topoisomerase I was carried out in the presence of Mg^{2+} while electrophoresis was in the absence of Mg^{2+}. When DNA was incubated with saturating amounts of TFIID, a portion of topoisomers was shifted to a more negative side. The linking number change was not observed if heat-inactivated TFIID was used or if the TATA sequence of Ad2 MLP was replaced by a *Hin*dIII linker. Under the conditions employed, the shift of topoisomers was barely detectable on relaxed circular DNA. These results suggest that the functional binding of TFIID to the Ad2 MLP accompanies underwinding of duplex DNA and that negative supercoiling of DNA facilitates the process.

Initiation of transcription by RNA polymerase II requires the formation of the preinitiation complex on a promoter. The first step is recognition of the TATA element by TFIID. Next, TFIIB binds to the TFIID-TATA element complex. These interactions mediate the stepwise assembly of TFIIE/TFIIF, and RNA polymerase II (*10, 11*). Each step involves specific protein-DNA and/or protein-protein interactions which are a prerequisite for the subsequent steps. The observed untwisting of DNA by TFIID might serve as a landmark for installation of TFIIB on the TFIID-TATA element complex, because TFIIB binds strongly to single-stranded DNA without recognizing any specific sequences on the DNA (*12*). The subsequent assembly of other general factors and RNA polymerase II seems to result in the "open complex" formation as a prelude to initiation.

It has been shown that sequence specific DNA binding proteins play an important role in the regulation of eukaryotic genes (see ref. *13* for a review). These proteins appear to exert their functions by modulating the utilization of promoter. However, the precise mechanisms of their action are not yet clear. We have shown here that negative supercoiling of DNA facilitates the underwinding of DNA induced by TFIID-TATA element interactions. It is possible that the process is also accelerated by the action of transcription regulatory factors.

SUMMARY

In vitro transcription was reconstituted with HeLa cell transcription factors and RNA polymerase II, which were essentially free from DNA topoisomerase activities. DNA templates with defined negative superhelical densities were tested for transcription activity. Transcription of the *B. mori* fibroin gene increases and plateaus from templates of increasing superhelicity and transcription from the adenovirus 2 major late promoter rises and then falls, while transcription of the *Drosophila hsp70* gene remains unchanged. Dissection of transcription into pre- and post-initiation steps using Sarkosyl reveals that formation of a preinitiation complex on the fibroin gene or the adenovirus 2 major late promoter is slow on relaxed DNA and accelerated by DNA superhelicity. Furthermore, additional experiments suggests that DNA supercoiling facilitates a functional binding of TFIID to the fibroin gene and adenovirus 2 major late promoter. Binding of bacterially expressed yeast TFIID to the adenovirus 2 major late promoter induces underwinding of duplex DNA. This conformation change is dependent on DNA superhelicity. It is most likely that the observed untwisting of DNA triggers an "open complex" formation prior to transcription initiation.

Acknowledgment
This work was supported in part by Grants-in-Aid for Scientific Research from the Ministry of Education, Science and Culture of Japan.

REFERENCES

1. M. Gellert: *Annu. Rev. Biochem.* **50**, 879–910 (1981).
2. J.C. Wang: *Annu. Rev. Biochem.* **54**, 665–697 (1985).
3. R.R. Sinden, J.O. Carlson, and D.E. Pettijohn: *Cell* **21**, 773–783 (1980).
4. S. Hirose, M. Tsuda, and Y. Suzuki: *J. Biol. Chem.* **260**, 10557–10562 (1985).
5. S. Hirose and Y. Suzuki: *Proc. Natl. Acad. Sci. U.S.A.* **85**, 718–722 (1988).
6. H. Tabuchi and S. Hirose: *J. Biol. Chem.* **263**, 15282–15287 (1988).
7. T. Ohta and S. Hirose: *Proc. Natl. Acad. Sci. U.S.A.* **87**, 5307–5311 (1990).
8. M. Mizutani, T. Ohta, H. Watanabe, H. Handa, and S. Hirose: *Proc. Natl. Acad. Sci. U.S.A.* **88**, 718–722 (1991).
9. M. Mizutani, K. Ura, and S. Hirose: *Nucl. Acids Res.* **19**, 2907–2911 (1991).
10. S. Buratowski, S. Hahn, L. Guarente, and P.A. Sharp: *Cell* **56**, 549–561 (1989).
11. M.W. Van Dyke, R.G. Roeder, and M. Sawadogo: *Science* **241**, 1335–1338 (1988).
12. D. Reinberg, M. Horikoshi, and R.G. Roeder: *J. Biol. Chem.* **262**, 3322–3330 (1987).
13. P. Johnson and S.L. McKnight: *Annu. Rev. Biochem.* **58**, 799–839 (1989).

Molecular Structure and Life (Ed. by Y. Kyogoku and Y. Nishimura), pp. 237–249, Japan Sci. Soc. Press, Tokyo/CRC Press, Boca Raton (1992)

Mechanism of Transcription Activation by cAMP Receptor Protein of *Escherichia coli*

Hiroji AIBA, Chisato USHIDA, Hideaki TAGAMI, and Tadashi MIYAZAKI

*Department of Molecular Biology, School of Science, Nagoya University**

The cAMP receptor protein (CRP) of *Escherichia coli*, also known as the catabolite gene activator protein (CAP), is a sequence specific DNA-binding protein which regulates transcription of a number of genes in response to carbon source limitation (*1–3*). When a carbon source such as glucose is limited in the culture medium, intracellular cAMP levels are raised, resulting in higher levels of the CRP-cAMP complex. When complexed with cAMP, CRP undergoes a conformational transition (*4*, *5*) and binds to specific sites within or near target promoters to activate or inhibit transcription. The CRP is a dimer of two identical subunits composed of 209 amino acids (*6–8*). The subunit has a two-domain structure. The larger N-terminal domain is responsible for cAMP binding and subunit-subunit interaction while the smaller C-terminal domain contains a helix-turn-helix motif involved in DNA binding (*4*, *5*, *9*, *10*).

The CRP-binding sites in various genes include variations of the consensus sequence that is 22 bp and exhibits a perfect two-fold symmetry (*11–14*). One of the most interesting features of CRP-dependent promoters is that the CRP-binding site is located at different positions

* Nagoya, Aichi 464, Japan

with respect to the transcription start site (*1–3*). Although direct contact between CRP and RNA polymerase bound to DNA has been believed to play an important role in transcription activation, the variety of locations of the CRP-binding site seemed to be incompatible with a simple protein-protein interaction model. To account for this, *lac* promoter mutants with altered spacing between the CRP-binding site and -35 sequence were characterized (*15*, *16*). The major conclusion of these previous studies was that the promoter with 11-bp insertion partially retained the ability to respond to CRP while the insertion of 5 bp eliminated the effect of CRP on transcription. Thus the activation by CRP seems to be dependent on the helical phase between two sites.

In the present work, we addressed questions of how strictly the stereospecific positioning of the CRP and RNA polymerase binding sites is needed for CRP action and how far the CRP site can be separated from the transcription start site and still carry out its function. We also isolated and characterized CRP mutants that bind DNA, but are unable to activate transcription. This allowed us to identify the site on CRP which is assumed to be involved in the interaction with RNA polymerase.

I. CONSTRUCTION OF SPACING MUTANTS

Plasmids used in the phasing experiment were all derived from pRCI3 (*17*) which contains an artificial promoter activated by CRP-cAMP. Plasmid pUT0 was constructed by removing one of the two CRP-binding sites from pRCI3. This plasmid carries promoter fused to the *lac* structural genes. Transcription from this promoter is strongly activated by CRP-cAMP. In pUT0 the center of the CRP-binding site is located at -71.5, and the distance between the downstream boundary of the CRP site and the upstream boundary of the -35 region is 25 bp. By inserting or deleting nucleotides in the spacer region, we constructed various spacing mutants (*18*). The insertions and deletions were designed to be either integral or half-integral turns of the DNA helix. The structural features of the spacing mutants are shown in Fig. 1. Plasmids pUT+21, pUT+11, pUT−10, and pUT−20 contain either insertions or deletions of integral turns of the DNA helix. These mutants are referred to as in-phase mutants, since the phasing of the CRP site and the -35 region is preserved as in the original. Plasmids pUT+25, pUT+15, pUT+4, pUT−5, and pUT−14 are out-of-phase mutants, since half-integral turns of the DNA helix were added or deleted in these mutants.

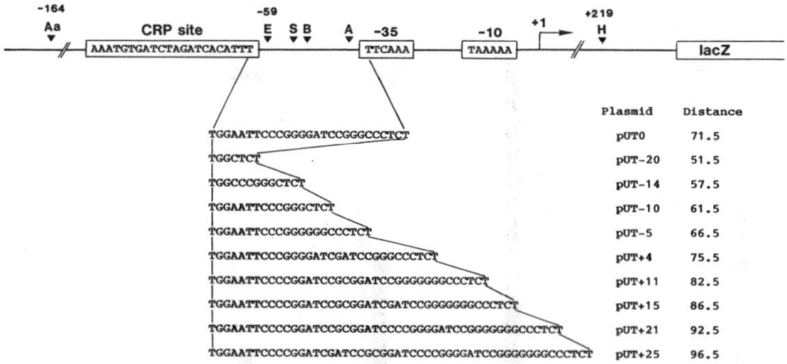

Fig. 1. Structures of spacing mutants. A series of promoters was derived from an artificial CRP-dependent promoter pRCI3 (*17*). The start site for transcription is indicated by an arrow designated +1. The CRP-binding site and −35 and −10 sequences are boxed. Abbreviations for restriction sites are: A, *Apa*I; Aa, *Aat*II; B, *Bam*HI; H. *Hin*dIII; E, *Eco*RI; S, *Sma*I. The distance is defined as the length (bp) between the center of the CRP site and the transcription start site.

II. BINDING OF CRP-cAMP TO SPACING MUTANTS

We previously found that CRP-cAMP strongly binds to the synthetic consensus CRP-binding site (*11*). A more quantitative binding assay by Ebright *et al.* (*13*) revealed that CRP exhibited a 450-fold higher affinity for the consensus CRP site than the *lac* CRP site. To examine the binding properties of CRP to the CRP sites in various spacing mutants, we performed a DNase I footprinting assay using restriction fragments containing the promoter region. With all of the mutant fragments used, a region of about 26-bp containing the CRP site was protected by CRP-cAMP from DNase I attack. The binding pattern and affinity of CRP to the CRP site are essentially the same among promoter mutants.

III. *IN VIVO* PROPERTIES OF SPACING MUTANTS

To examine the effect of the spacer length on CRP activation, we introduced plasmid constructs into crp^+ and Δcrp cells, and measured the expression of the *lacZ* gene in cells. The transcription of the promoter in the parental plasmid pUT0 is highly dependent on CRP. Inserting or deleting DNA of non-integral helical turns between the CRP site and the −35 region totally eliminated the β-galactosidase activity in crp^+ cells, while the 11-bp and 21-bp insertions gave β-galactosidase levels of 50% and 20% of the parent promoter, respectively (Fig. 2). Interestingly, the promoter activity markedly increased with the introduction of 10-bp

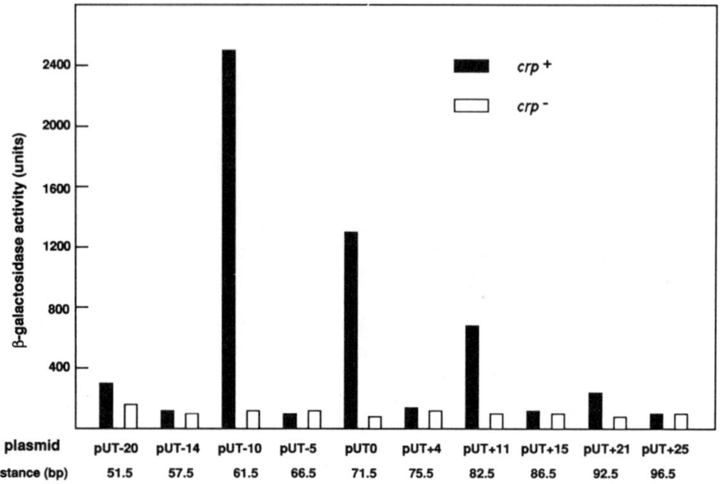

Fig. 2. Promoter activities of spacing mutants. The β-galactosidase assay was performed with strains TB100 (*Δlac crp*$^+$ *cya*$^+$) and TB102 (*Δlac Δcrp cya*$^+$) harboring each plasmid. The activity is expressed in Miller units.

deletion (180% of the original activity); however, the 20-bp deletion resulted in a considerable decrease in the activity. In *Δcrp* cells all the spacing mutants exhibited low levels of β-galactosidase activity. Thus the promoter activities of in-phase mutants are dependent on CRP as in the parental plasmid pUT0. These results clearly indicate that transcription activation by CRP is strictly dependent on the helical phase between the CRP site and the −35 sequence.

IV. S1 ANALYSIS OF PROMOTER ACTIVITY

Although the −35 and −10 sequences of the spacing mutants are the same as those of pUT0, we cannot exclude the possibility that insertions or deletions of nucleotides in the spacer region alter the start site for transcription in the original promoter. To examine this and to characterize further promoter activity, we performed a quantitative S1 assay. Cellular RNAs from *crp*$^+$ and *Δcrp* cells harboring plasmids were hybridized to the *Eco*RI-*Hin*dIII fragment, derived from pUT0 and 32P-labeled at its *Hin*dIII 5′ end. In this assay RNA from cells harboring the parental plasmid pUT0 gave an S1 resistant DNA band of about 220 bases. The same size S1 resistant DNA bands were also produced by RNAs from all the spacing mutants. This indicates that the start sites for transcription in mutant promoters are the same as that of pUT0. In addition, the variation of the intensity of the S1 resistant DNA bands clearly indicates that the promoter activity varies periodically depending on the helical phase between the binding sites for CRP and RNA polymerase. The results of S1 assay are fully consistent with those of β-galactosidase assay, indicating again that the activation of transcription by CRP is strictly dependent on stereospecific alignment between the CRP and RNA polymerase binding sites.

V. POSITIVE CONTROL MUTANTS OF CRP

In an attempt to locate sites on CRP that may be involved in the interaction with RNA polymerase, we have isolated and characterized mutant CRPs that bind to DNA, but are incapable of activating transcription. Such mutations are referred to as positive control (pc) mutants (*19*). To isolate pc mutants of CRP, we used strain TP7839 (*Δcrp*) harboring plasmid pPxSL604 (*11*) as a tester strain. Plasmid pPxSL604 carries a *Px-lac* fusion gene which possesses a consensus CRP-binding sequence between the promoter and the *lacZ* gene. The expression of this fusion gene is under the negative control of CRP-cAMP. In other words, the gene is expressed in the absence of CRP-cAMP while it is repressed in the presence of CRP-cAMP. When CRP-cAMP binds to this site, the transcription is strongly repressed. TP7839 harboring pPxSL604 displays a Lac$^+$ phenotype, since the *Px-lac* fusion gene is expressed normally. When the cells are transformed with a second plasmid containing the *crp* plasmid (pAcrp), the *Px-lac* fusion gene is turned off but the *lac* gene on the chromosome is activated because of the presence of CRP-cAMP. Therefore, TP7839 harboring both pPxSL604 and pAcrp is also Lac$^+$. If mutant CRP is made in cells that can bind to the CRP site but has lost the ability to activate transcription, however, the cells are expected to show a Lac$^-$ phenotype. We first mutagenized pAcrp by UV irradiation or hydroxylamine treatment *in vitro* and then transformed it to a *Δcrp* strain TP7839 harboring pPxSL604. Although most transformants displayed Lac$^+$ phenotype on MacConkey-lactose plates, several transformants revealed a Lac$^-$ phenotype. Plasmid DNA from 8 independently isolated *Lac*$^-$ colonies was transformed again to TP7839/pPxSL604. Of these 8 plasmids, 5 conferred a Lac$^-$ phenotype stably to the tester strain. We have determined the nucleotide sequences of 5 pc mutants and the sequence analyses of the mutants revealed three different mutations. The amino acid substitutions in the three mutants were mapped to a distinct

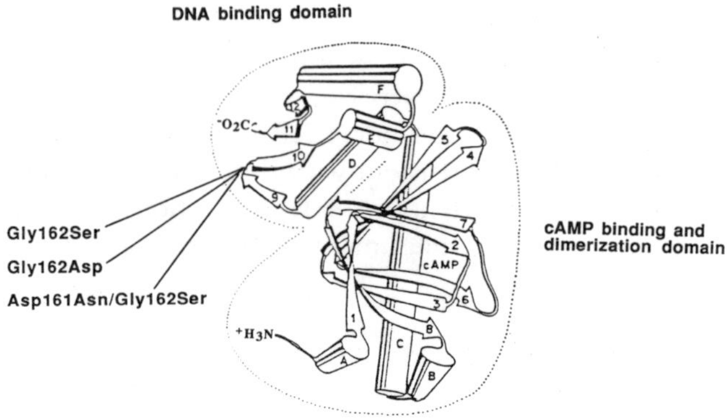

Fig. 3. Amino acid changes in positive control mutants. The location of amino acids substituted in the three CRP mutants is shown on the CRP tertiary structure. All of the mutations are mapped to the turn region between β9 and β10 sheets. Schematic drawing of a CRP monomer is taken from McKay *et al.* (*10*).

region of CRP: Gly162Ser, Gly162Asp, and Asp161Asn/Gly162Ser. This position is located between the β rolls 9 and 10 in the tertiary structure of CRP (Fig. 3).

VI. PROPERTIES OF CRP MUTANTS

To characterize quantitatively the properties of the CRP mutants *in vivo*, we measured the transcription activation of several promoters by CRP mutants. First, we introduced plasmids carrying mutant *crp* alleles into strain TP2139 (*Δlac Δcrp*) harboring pPxSL604 and measured the β-galactosidase activities to test their ability to bind to CRP sites. This promoter is active in the absence of CRP, but its activity is strongly inhibited by the binding of CRP-cAMP. Therefore, CRP-dependent inhibition of expression from this promoter can be used to monitor CRP binding to target sites *in vivo*. The results showed that all of the CRP mutants repressed as efficiently as wild-type CRP did, indicating that the mutant CRPs retain the ability to bind to the CRP site *in vivo* (Table I). Next, we tested the ability of CRP mutants to activate transcription from CRP-dependent promoters by introducing the *crp* alleles into strain TP2139 harboring a *lacP-lac* fusion plasmid pTM03. As shown in Table I, the wild-type CRP can activate transcription very efficiently from the *lac* promoter, while all of the mutant CRPs essentially failed to stimulate the *lac* transcription. We also tested the ability of CRP mutants to

TABLE I
Properties of Positive Control Mutants of CRP *In Vivo*

CRP (plasmid)	Activation of *lacP*[a)]		Activation of *PI3*[b)]		Repression of *Px*[c)]	
	β-Gal Activity	Ratio	β-Gal Activity	Ratio	β-Gal Activity	Ratio
None (pACYC184)	71	1.0	235	1.0	1,945	1.00
wt (pAcrp-wt)	2,456	34.6	2,683	13.3	113	0.06
G162S (pAcrp152)	137	1.9	198	0.8	150	0.08
G162D (pAcrp190)	84	1.2	177	0.8	195	0.10
D161N/G162S (pAcrp191)	95	1.3	137	0.6	156	0.08

Three promoters/strains were used to examine the action of CRP *in vivo*: (a) the *lac* promoter on chromosome/TP7839 (*Δcrp*); (b) an artificial CRP-dependent promoter *PI3* on plasmid pUT-10/TP2139 (*Δcrp Δlac*) harboring pUT-10; (c) *Px* on plasmid pPxSL62/TP2139 harboring pPxSL62. The pACYC184 derivatives carrying the *crp* alleles were transformed into tester strains. Each strain was grown in LB medium at 37°C to $A_{600}=0.8$ and the β-galactosidase activity was determined. The activity is expressed in Miller units.

stimulate transcription from an artificial CRP-dependent promoter that contains a consensus CRP-binding site. TP2139 harboring a plasmid pUT0 carrying this promoter was transformed by pAcrp derivatives carrying each of the mutated *crp* genes. The results showed that all of the mutant CRPs completely lost the ability to stimulate transcription from this promoter (Table I). Thus, the mutant CRPs are defective in transcription activation of the type I CRP-dependent promoter in which the CRP-binding site is located upstream of the −35 sequence (*17*).

Are these mutant CRPs also defective in activating of type II CRP-dependent promoters in which the CRP site overlaps the −35 sequence? To examine this we conducted similar assays using pBRP4 (*20*) and the divergent promoter of the *crp* region (*21*) fused to *lacZ* on plasmids, *P4-lac* and *Pd-lac* fusions, respectively. The transcription from the two promoters is totally dependent on CRP-cAMP and the CRP sites in these promoters overlaps the −35 sequence. We found that the mutant CRPs more or less retain the ability to activate the type II CRP-dependent promoters. For example, Gly162Ser activates transcription from *Pd-lac* and *P4-lac* by 80% and 40%, respectively, of the level with wild-type CRP. Two other mutants were less effective in the activation of transcription from both promoters.

Gel binding assays were carried out to examine the interactions between mutant CRPs and CRP-binding sequences using partially purified mutant proteins. All of the mutant CRPs were shown to bind to the DNA fragment containing the CRP-binding sites *in vitro*. However, a more quantitative assay showed that the binding affinity of the mutant

CRPs to the CRP sites was significantly lower than that of the wild-type CRP.

VII. HELICAL PHASE DEPENDENT ACTION OF CRP

A prevailing view concerning the mechanism of transcriptional activation both in prokaryotes and eukaryotes is that the activation is mediated by direct interaction between activator protein(s) and transcriptional machinery (*22–25*). In the case of CRP, several lines of evidence suggest that CRP activates transcription by contacting RNA polymerase. First, fluorescence, ultracentrifugation, and immunological studies suggest the interaction between two proteins (*26–29*). Second, cooperative binding to the *lac* promoter region of CRP and RNA polymerase is demonstrated by protection experiments such as DNase I footprinting (*30*, *31*). Third, the importance of the CRP-RNA polymerase interaction is supported by the isolation of CRP mutants that lack the activation, but are able to bind to DNA (*19*, *32*, *53*).

Further evidence supporting CRP-RNA polymerase interaction comes from phasing experiments. Initially, Mandecki and Caruthers (*15*) characterized the *lac* promoter mutants with varying lengths of the spacer between the CRP-binding site and the −35 region. They found that the insertion of 5 bp between the two sites eliminated CRP activation, while the activation was partially restored with 11-bp insert. More recently, Straney *et al.* (*16*) have shown using similar spacing mutants that stabilization of CRP binding to the *lac* DNA by RNA polymerase occurs in parallel with CRP activation. These data support the view that the interaction between CRP and RNA polymerase bound on the same side of the DNA helix is important for CRP action. However, in earlier phasing experiments the variation in the spacer length was limited to within one turn of the DNA helix. In addition, the start sites for transcription in spacer mutants were not analyzed.

In the present work we carried out a more extensive phasing experiment using a synthetic CRP-dependent promoter. We constructed a set of spacing mutants in which the distance between the CRP site and the −35 region varied over 4 turns of DNA helix. Then, we analyzed the transcriptional activity of spacing mutants *in vivo* by β-galactosidase and S1 assays. All in-phase mutants, which preserve the integral helical spacing between the CRP site and the −35 region, have been shown to retain to some degree the ability to be activated by CRP, whereas out-of-phase mutants, which have insertion or deletion of non-integral helical turns, can no longer respond to CRP. The observation that the periodicity of the stimulation by CRP is around 10.5 bp is consistent with those of the previous phasing experiments on the *lac* promoter. Since we observed no significant differences in binding of CRP-cAMP to DNA among the spacing mutants, the periodical variation of promoter activity strongly suggests that the positioning of the CRP and RNA polymerase binding sites on the same face of DNA helix is strictly required for CRP action (Fig. 4). The helical phase dependence of CRP action supports the model that CRP activates transcription by contact with RNA polymerase. The importance of stereospecific positioning of the binding sites for an activator and RNA polymerase has also been shown in several other promoters by analogue phasing experiments (*33–36*). Our data indicate that CRP can most efficiently activate transcription when the center of its binding site is positioned 61.5 bp upstream from the start site. This location is exactly the same as that in *lac* (*37*) and *tna* (*38*). In addition, it has been shown that the placement of the CRP site at −61.5 in the *lac-gal* hybrid promoter is also very effective for CRP activation (*39*). Thus, in several different promoters CRP can activate transcription very efficiently at −61.5. This position is apparently optimum for CRP to interact with RNA polymerase in type I CRP-dependent promoter in which the CRP site is separated from the −35 region. The activation by

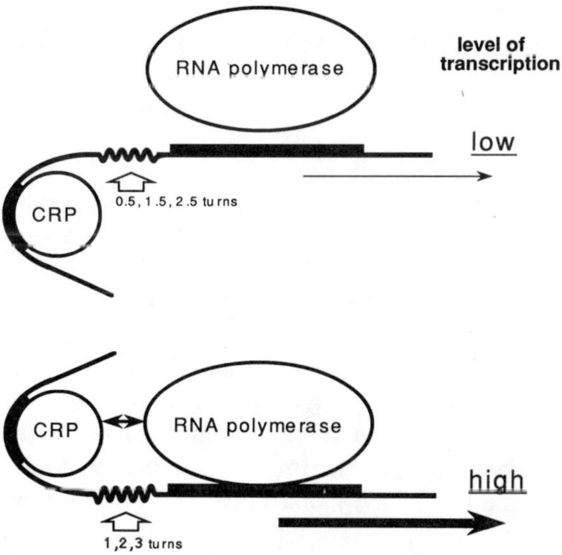

Fig. 4. Model for CRP action. Stereospecific positioning of the CRP and RNA polymerase binding sites is needed for the transcription activation by CRP.

CRP decreases rapidly with the deviation from the optimum spacer length. However, we still observed a significant activation when the CRP site was placed 92.5 bp upstream from the start site. In naturally occurring promoters, where CRP acts as the sole activator, the greatest distance between the CRP site and the transcription start site is 70.5 bp, as seen in *malT* (*40*). The CRP sites are located further upstream in *araBAD* (*41*) and *malE-malK* (*42*) operons. In these cases, however, additional proteins are required for CRP action (*43*, *44*). Recently, it has been found that the CRP site at *papB* is separated more than 200 bp from the start site (*45*). It remains to be seen if CRP acts alone or with additional factors in this operon.

The contact between two proteins at variable distances should require structural distortions in DNA and/or proteins. We have observed that CRP induces a bending upon binding to the promoter fragments derived from pUT series plasmids as in the case of the *lac* promoter. However, this bending alone is probably not sufficient to account for the long range CRP-RNA polymerase interaction, since the CRP-induced bend is thought to occur around the center of the CRP site (*46*, *47*). Additional bending or looping out in the spacer region would be needed to accommodate the protein-protein interaction. Loop formation involving an activator protein and σ54 form of RNA polymerase has now been well established at several nitrogen-regulated promoters in bacteria (*48*, *49*). On the other hand, possible structural distortions in DNA mediated by CRP-RNA polymerase interaction remain to be studied.

VIII. ACTIVATION SURFACE ON CRP

The isolation of pc mutants of CRP further supports the view that CRP stimulates transcription by contacting RNA polymerase. The sequence determination defined the sites in which amino acid substitutions occur. These sites are mapped in the very specific region between β9 and β10 sheets of the tertiary structure of CRP-cAMP complex. Bell *et al.* (*32*), and Eschenlauer and Reznikoff (*53*) isolated several pc mutants of CRP that have an amino acid substitution either at positions 159 or 162. Thus several pc mutations isolated independently all lie in the same region. Although this region is close to the DNA-binding domain, it is clearly separated from the helix-turn-helix motif. We propose that this region is involved in the contact with RNA polymerase in type I CRP-dependent promoter, since it projects out from the surface of CRP. However, it is apparent that mutations in this region also affect the interaction between

CRP and CRP site, since the mutant CRPs exhibit a decreased affinity to CRP-binding sites. It is noteworthy that our pc mutants to some degree retain the ability to activate transcription at so-called type II CRP-dependent promoter. This suggests that CRP may have other sites which are involved in the interaction with RNA polymerase. In fact, Irwin and Ptashne isolated several pc mutants which have amino acid substitutions in the first helix (αE) of the helix-turn-helix motif (*19*). The αE mutations decrease activation at the *gal* promoter (type II) more strongly than at the *lac* or *malT* promoters (type I). Thus CRP appears to possess two surfaces which are involved in transcriptional activation. Gaston *et al.* (*50*) have also characterized a set of mutants with varying lengths between the CRP site and the -10 region of *melR* promoter. They showed that the activation by CRP required a stringent spacing between the two sites. In addition, they found that the distances at which CRP activates transcription correspond exactly to those at the *galP1* (type II) and *lacP1* (type I), proposing a two-position model for the interaction of CRP and RNA polymerase. Igarashi and Ishihama (*51*) found that RNA polymerases containing C-terminally truncated α subunits can no longer respond to transcription activation by CRP-cAMP. This finding supports strongly from the RNA polymerase side that transcriptional activation is mediated by direct interaction between CRP and polymerase. Moreover, it has been shown that the same mutant RNA polymerases can activate type II CRP-dependent promoters (*52*). This result is consistent with a two position model for the interaction between CRP and RNA polymerase.

SUMMARY

We show evidence which suggests that the mechanism of transcriptional activation by *E. coli* CRP is mediated by direct interaction between CRP and RNA polymerase. First, a series of spacing mutants derived from an artificial CRP-dependent promoter that have varying distances between the CRP site and the -35 region was constructed. The effect of the spacer length on transcription activation by CRP-cAMP has been examined *in vivo* by β-galactosidase and S1 assays. The results clearly demonstrate that the activation by CRP is strictly dependent on the helical phase between the CRP site and the -35 region. The activation by CRP occurs most efficiently when the CRP site is located 61.5 bp upstream from the transcription start site. In a second approach, we isolated and characterized positive control CRP mutants that bind DNA, but are unable to activate transcription. Five independent mutants were isolated and sequenced to obtain three different mutants. The mutations are located at positions 162 and 161 which are just between the $\beta 9$ and $\beta 10$ sheets of the CRP tertiary structure. This region appears to protrude from the surface of the protein. We propose that this region would be involved in transcription activation by coming in contact with RNA polymerase and the mutations may affect this protein-protein interaction.

Acknowledgment
This work was supported in part by Special Project Research Funds from the Ministry of Education, Science, Culture of Japan.

REFERENCES

1. A. Ullmann and A. Danchin: *Adv. Cyclic Nucleotide Res.* **15**, 1–53 (1983).
2. B. de Crombrugghe, S. Busby, and H. Buc: *Science* **224**, 831–838 (1984).
3. S. Busby: In "Regulation of Gene Expression" (Ed. by I. Booth and C. Higgins), pp. 51–77, Cambridge University Press, Cambridge (1986).
4. J.S. Krakow and I. Pastan: *Proc. Natl. Acad. Sci. U.S.A.* **70**, 2529–2533 (1973).
5. E. Eilen, C. Pampeno, and J.S. Krakow: *Biochemistry* **17**, 2469–2473 (1978).
6. W.B. Anderson, A.B. Schneider, M. Emmer, R.L. Perlman, and I. Pastan: *J. Biol. Chem.* **246**, 5929–5937 (1971).
7. H. Aiba, S. Fujimoto, and N. Ozaki: *Nucl. Acids Res.* **10**, 1345–1361 (1982).
8. P. Cossart and B. Gicquel-Sanzey: *Nucl. Acids Res.* **10**, 1363–1378 (1982).
9. H. Aiba and J.S. Krakow: *Biochemistry* **20**, 4774–4780 (1981).
10. D.B. McKay and T.A. Steitz: *Nature* **290**, 744–749 (1981).
11. T. Morita, K. Shigesada, F. Kimizuka, and H. Aiba: *Nucl. Acids Res.* **16**, 7315–7332 (1988).
12. O.G. Berg and P.H. von Hippel: *J. Mol. Biol.* **200**, 709–723 (1988)
13. R.H. Ebright, Y.W. Ebright, and A. Gunasekera: *Nucl. Acids Res.* **17**, 10295–10305 (1989).
14. K. Gaston, A. Kolb, and S. Busby: *Biochem. J.* **261**, 649–653 (1989).
15. W. Mandecki and M.H. Caruthers: *Gene* **31**, 263–267 (1984).
16. D.C. Straney, S.B. Straney, and D.M. Crothers: *J. Mol. Biol.* **206**, 41–57 (1989).
17. H. Aiba, A. Hanamura, and T. Tobe: *Gene* **85**, 91–97 (1989).
18. C. Ushida and H. Aiba: *Nucl. Acids Res.* **18**, 6325–6330 (1990).
19. N. Irwin and M. Ptashne: *Proc. Natl. Acad. Sci. U.S.A.* **84**, 8315–8319 (1987).
20. C. Queen and M. Rosenberg: *Nucl. Acids Res.* **9**, 3365–3377 (1981).
21. A. Hanamura and H. Aiba: *Nucl. Acids Res.* **19**, 4413–4419 (1991).
22. K. Gaston, A. Bell, A. Kolb, H. Buc, and S. Busby: *Cell* **62**, 733–743 (1990).
23. M. Ptashne: *Nature* **322**, 697–701 (1986).
24. M. Ptashne: *Nature* **335**, 683–689 (1988).
25. J.D. Gralla: *Cell* **57**, 193–195 (1989).
26. F.Y.-H. Wu, K. Nath, and C.-W. Wu: *Biochemistry* **13**, 2567–2572 (1974).

27. M. Pinkney and J.G. Hoggett: *Biochem. J.* **250**, 897–902 (1988).
28. B. Blazy, M. Takahashi, and A. Baudras: *Mol. Biol. Rep.* **6**, 39–43 (1980).
29. W. Stender: *Biochem. Biophys. Res. Commun.* **96**, 320–325 (1980).
30. A. Spassky, S. Busby, and H. Buc: *EMBO J.* **3**, 43–50 (1984).
31. Y.L. Ren, S. Garges, S. Adhya, and J.S. Krakow: *Proc. Natl. Acad. Sci. U.S.A.* **85**, 4138–4142 (1988).
32. A. Bell, K. Gaston, K. Williams, A. Kolb, H. Buc, S. Minchin, J. Williams, and S. Busby: *Nucl. Acids Res.* **18**, 7243–7250 (1990).
33. M. Buck, W. Cannon, and J. Woodcock: *Mol. Microbiol.* **1**, 243–249 (1987).
34. S. Maeda, Y. Ozawa, T. Mizuno, and S. Mizushima: *J. Mol. Biol.* **202**, 433–441 (1988).
35. S.D. Minchin, S. Austin, and R.A. Dixon: *EMBO J.* **8**, 3491–3499 (1989).
36. L.J. Reitzer, B. Movas, and B. Magasanik: *J. Bacteriol.* **171**, 5512–5522 (1989).
37. R.C. Dickerson, J. Abelson, W.M. Barnes, and W. Reznikoff: *Science* **187**, 27–35 (1975).
38. M.C. Deely and C. Yanofsky: *J. Bacteriol.* **151**, 942–951 (1982).
39. L. Bracco, D. Kotlarz, A. Kolb, S. Diekmann, and H. Buc: *EMBO J.* **8**, 4289–4296 (1989).
40. C. Chapon and A. Kolb: *J. Bacteriol.* **156**, 1135–1143 (1983).
41. N.L. Lee, W.O. Gielow, and R.G. Wallace: *Proc. Natl. Acad Sci. U.S.A.* **78**, 752–756 (1981).
42. H. Bedouelle: *J. Mol. Biol.* **170**, 861–882 (1983).
43. N.L. Lee, C. Franklyn, and E.P. Hamilton: *Proc. Natl. Acad. Sci. U.S.A.* **84**, 8814–8818 (1987).
44. O. Raibaud, D. Vidal-Ingigliardi, and E. Richet: *J. Mol. Biol.* **205**, 471–485 (1989).
45. M. Gransson, K. Forsman, P. Nilsson, and B.E. Uhlin: *Mol. Microbiol.* **3**, 1557–1565 (1989).
46. H.-M. Wu and D. Crothers: *Nature* **308**, 509–513 (1984).
47. H.-N. Liu-Johnson, M.R. Gartenberg, and D.M. Crothers: *Cell* **47**, 995–1005 (1986).
48. W. Su, S. Porter, S. Kustu, and H. Echols: *Proc. Natl. Acad. Sci. U.S.A.* **87**, 5504–5508 (1990).
49. T.R. Hoover, E. Santero, S. Porter, and S. Kustu: *Cell* **63**, 11–22 (1990).
50. K. Gaston, A. Bell, A. Kolb, H. Buc, and S. Busby: *Cell* **62**, 733–743 (1990).
51. K. Igarashi and A. Ishihama: *Cell* **65**, 1015–1022 (1991).
52. K. Igarashi, A. Hanamura, K. Makino, H. Aiba, H. Aiba, T, Mizuno, A. Nakata, and A. Ishihama: *Proc. Natl. Acad. Sci. U.S.A.* **88**, 8598 8962 (1991).
53. A.C. Eschenlauer and W.S. Reznikoff: *J. Bacteriol.* **173**, 5024–5029 (1991).

Molecular Structure and Life (Ed. by Y. Kyogoku and Y. Nishimura), pp. 251–260, Japan Sci. Soc. Press, Tokyo/CRC Press, Boca Raton (1992)

A DNA-loop Model for Transcriptional Activation of the Xylene-metabolizing Genes on *Pseudomonas* TOL Plasmid

Atsushi NAKAZAWA,*1 Sachiye INOUYE,*1 Manabu GOMADA,*1 Hiromasa IMAISHI,*1 and Teruko NAKAZAWA*2

*Department of Biochemistry**1 *and Department of Microbiology,**2 *Yamaguchi University School of Medicine*

A strain of *Pseudomonas putida* called mt-2 can catabolize xylene because a series of enzymes that degrade this compound is induced when the cells are exposed to the compound. A methyl group of xylene is oxidized in the compound making it toluate, and then the benzene ring is opened, leading to further conversion to acetate, pyruvate and acetoacetate which enter the TCA cycle. At least 12 enzymes and proteins are involved in this pathway, and all are encoded by the TOL plasmid, a degradative plasmid of *Pseudomonas* (*1*).

Several unique molecular mechanisms are involved in this induction system. The present review discusses the mechanism of transcriptional activation through the upstream regulatory sequence (URS) in the two promoters in the xylene-metabolizing pathway.

I. INDUCTION WITH CASCADE AMPLIFICATION

The genes for xylene degradation form two operons (Fig. 1; *2*, *3*). The first operon (*OP1*) is responsible for the upper metabolic pathway from

*1,*2 Ube, Yamaguchi 755, Japan

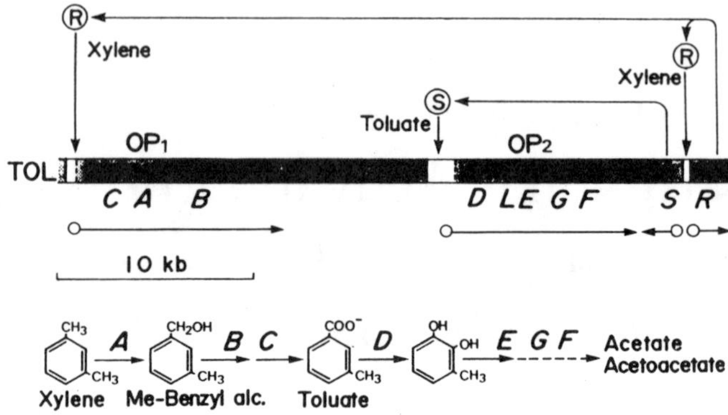

Fig. 1. Organization and expression of xylene-metabolizing genes on the *Pseudomonas* TOL plasmid.

xylene to toluate, and the second (*OP2*) is required for the lower catabolic pathway from the aromatic carboxylic acid to the final products. When xylene, a substrate of the entire pathway, is present in the environment, the product of a regulatory gene, *xylR* (XylR, 67 kDa), which is presumably bound by xylene, activates both *OP1* and another regulatory gene, *xylS*. Thus, enzymes encoded by *OP1* are induced, resulting in the conversion of xylene to toluate. Simultaneously, the synthesis of the *xylS* gene product (XylS, 37 kDa) in large quantity is induced, activating *OP2*, which leads to complete catabolism of toluate. In the latter process, toluate, which is a metabolic intermediate of xylene and the substrate of the lower catabolic pathway, binds to XylS protein to greatly enhance activation of *OP2* as an inducer as well (*4–6*). Thus the gene expression in xylene metabolism is regarded as a cascade amplification system involving induction of a regulatory protein in response to the pathway substrate. The entire regulatory system of xylene metabolism is based on the expression of the *xylR* gene, which is controlled by autogenous repression (*4*).

II. NtrA-DEPENDENT TRANSCRIPTION

Among four transcriptional units in the xylene degradation system, transcriptions of *OP1* and the *xylS* gene are subject to a common regulation: both are activated by XylR in the presence of xylene. The nucleotide sequences of the regions around transcription start points of *OP1* and the *xylS* gene share homology, which in turn have a close similarity to the nitrogen-regulated (*ntr*) and the nitrogen-fixation (*nif*)

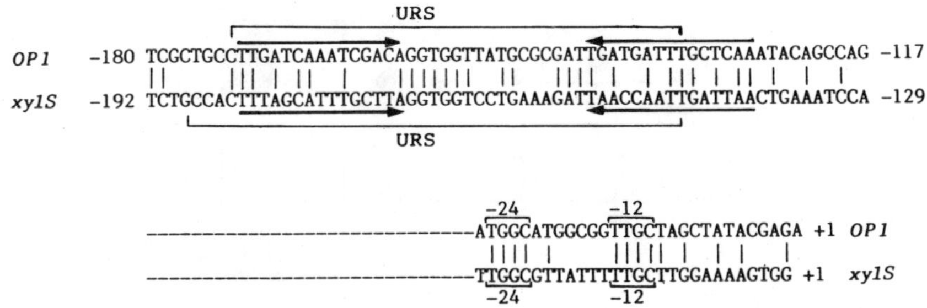

Fig. 2. Sequence comparison of the regions upstream from transcription start points of *OP1* and the *xylS* gene. These start points are designated by +1. Identical bases are connected by vertical lines. Horizontal arrows indicate palindromes in URSs.

promoters (Fig. 2). The consensus promoter sequences TGGC (−24) and TTGC (−12) are known to be recognized by RNA polymerase containing a sigma factor NtrA, σ^{54} (*7*). Using an *ntrA* mutant strain of *Escherichia coli*, it was shown that activation of both *OP1* and the *xylS* gene is dependent on the *ntrA* gene (*4*, *6*, *8*). We and others cloned the *ntrA* gene of *P. putida* and constructed a *P. putida ntrA* mutant by insertional inactivation (*9–11*). Using the mutant strain, requirement of the intact *ntrA* gene for activation of both *OP1* and the *xylS* gene in *P. putida* was demonstrated.

III. UPSTREAM REGULATORY SEQUENCES

All the NtrA-dependent promoters so far characterized require activator proteins which bind to DNA segments termed upstream regulatory sequences (URS) and locate about 100 bp upstream from the promoter (*12*). We have demonstrated the presence of URS in *OP1* and the *xylS* gene. Deletion analysis of the upstream regions revealed that the segment between −133 and −174 of *OP1* (*13*), and the segment between −145 and −189 of the *xylS* gene are required for efficient XylR-mediated induction by xylene (*14*).

Palindromic structures were recognized in URSs of *OP1* and the *xylS* gene (Fig. 2). These sequences that span about 40 bp were thought to serve as the binding sites for an oligomeric form of XylR protein, which has recently been verified by *in vitro* binding analysis (*15*).

IV. ENHANCER-LIKE PROPERTIES

The URS functions in an inverse orientation (Fig. 3). When the URS-

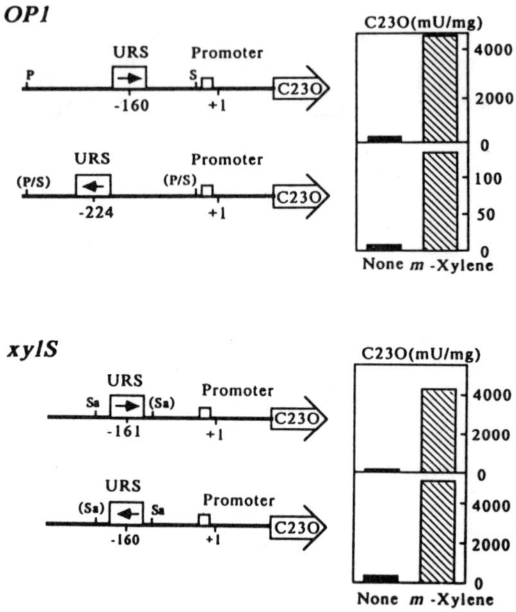

Fig. 3. Effect of orientation of URS on promoter activity of *OP1* and the *xylS* gene. For monitoring the promoter activity, expression plasmids were constructed in which catechol 2, 3-dioxygenase (C23O)-encoding sequence was ligated after the promoter regions of *OP1* and the *xylS*. C23O activity was measured with extracts of *E. coli* cells harboring each of these plasmids and the *xylR* plasmid which were cultured in the presence and absence of xylene. The original orientation of URS is shown by an arrow.

TABLE I
Effect of Long Insertions on *OP1* Promoter Activity

Plasmid	Insert (bp)	C23O (mU/mg)		Induction ratio (None / Xylene)
		None	Xylene	
pTS424	11	100	1,190	12
pTS468	200	3	220	74
pTS461	500	3	129	43
pTS459	1,200	13	97	7

Various length DNA segments were inserted between URS and the promoter of *OP1* after which the C23O-encoding sequence was ligated on an expression plasmid. *E. coli* cells carrying each of the plasmids and the *xylR* plasmid were cultured in the presence and absence of xylene.

containing segment of *OP1* was reversed, induced versus uninduced levels of gene expression (induction ratio) did not change, although the maximal activation levels decreased to 10% that of the original orientation (*13*). When the URS of the *xylS* gene was reversed, neither the

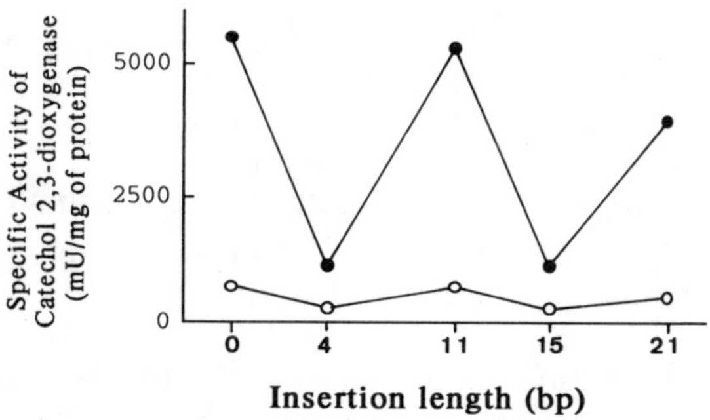

Fig. 4. Helical periodicity in induced expression of *OP1*. Short insertions (0, 4-, 11-, 15-, and 21-bp) were introduced between URS and the promoter of *OP1*, and the promoter activity was monitored by measuring C23O after incubating the cells in the presence (filled circles) and absence (open circles) of xylene.

maximal activation nor the induction ratio changed significantly (*14*).

When the various lengths of DNA fragments were inserted in the region between URS and the promoter of each of *OP1* and the *xylS* gene, the induction ratio did not change drastically, while the maximal activation decreased. Even with a 1.2-kb insertion, xylene induction of *OP1* was observed (Table I; *13*). Similar experiments showed that the *xylS* URS can also be displaced by 3.9 kb in the 5′-upstream region of the *xylS* promoter (*14*). These features of URSs of *OP1* and the *xylS* gene resemble those of eukaryotic enhancer sequences.

V. HELICAL PERIODICITY IN ACTIVATION

Insertion of shorter DNA segments up to 21 bp between URS and promoter in *OP1* brought about two clearly distinguishable states. Insertion of multiples of one turn of DNA helix resulted in maximal activation, while low activation was obtained upon insertion of a half turn of the helix (Fig. 4; *13*). Similar experiments showed that activation of *xylS* expression is also dependent on DNA helical periodicity of the insertion length between URS and the promoter of the *xylS* gene (*14*). These observations indicate that the URS must be strictly positioned with respect to the face of the DNA molecule along the long axis.

VI. TRANSCRIPTIONAL PROTEINS

The XylR protein is thought to bind to the URS through the C-terminal region which contains a helix-turn-helix motif (*16*). The central region of this protein is regarded as the site for interaction with NtrA-containing RNA polymerase, since this region is the most conserved among this kind of protein (*17*). We have recently found that autogenous repression of the *xylR* gene is abolished in the NtrA-minus background (*14*). Since the binding site of XylR in this repression is also used as the URS of the *xylS* gene which is divergently transcribed from the *xylR* gene, the above finding suggests that binding of XylR to the URS is dependent on NtrA protein.

NtrA protein is also thought to be a DNA-binding protein, since the region near the C-terminus contains a helix-turn-helix motif (*18*). The central region of NtrA protein, which has a sequence homologous with σ^{70} and β' subunit of RNA polymerase, probably plays a role in protein-protein interaction with subunits of core RNA polymerase (*19*).

VII. DNA-LOOP MODEL

Taking into account the helical periodicity for activation together with the considerations for protein-protein and protein-DNA interactions, we

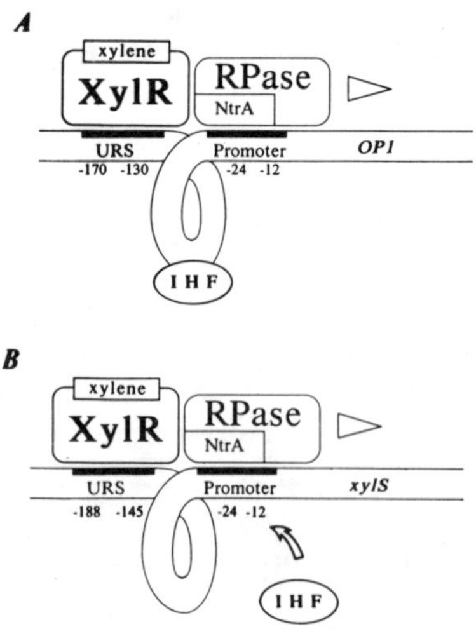

Fig. 5. A model for activation of *OP1* and the *xylS* promoters by XylR protein in the presence of xylene. XylR protein bound by xylene interacts with URS, while NtrA-containing RNA polymerase (RPase) binds to the promoter. The two proteins interact with each other through a DNA-loop formation to facilitate transcription (open arrows). IHF binds to the intermediate portion between URS and the promoter of OP1 (A), whereas it binds to the promoter of the *xylS* gene (B).

previously proposed a model for activation of *OP1* by XylR protein in the presence of inducer (Fig. 5). XylR protein bound by xylene interacts with the URS, while NtrA-containing RNA polymerase binds to the promoter sequence. Through the DNA-loop formation between the two sequences, XylR protein can interact with NtrA-containing RNA polymerase to form a quarternary complex among URS, XylR, NtrA-containing RNA polymerase, and the promoter sequence. This type of complex should be prerequisite to transcription initiation from *OP1*. The *xylS* gene has URS and the promoter at almost the same locations as in *OP1*. The same proteins are involved in activation of the *xylS* gene, and helical periodicity for activation was observed. Therefore, a similar DNA-loop mechanism would also be operating in the *xylS* gene expression.

Models involving such a quaternary complex with a DNA loop were proposed for activation of the *nifH* genes of *Klebsiella pneumoniae* and *Rhizobium meliloti* by NifA protein (*20*) and the *glnA* gene of *Salmonella typhimurium* (*21*). It was shown that NtrC protein in the complex hydrolyzes ATP to catalyze an open transcription initiation complex (*22*). The ATP-binding sequence homologous to that in NtrC protein is found in the center of XylR protein (*16*). Therefore, a similar mechanism involving ATP hydrolysis would be applicable to the *OP1* and *xylS* activation systems.

VIII. PROTEIN-PROTEIN INTERACTION

It was previously reported that the *OP1* promoter was activated not only by XylR protein but also by NtrC and NifA proteins in *E. coli* under nitrogen-limiting conditions (*8*). However, the identified DNA sequences for binding of NtrC protein and NifA protein (*7*) are not present in the upstream regions of *OP1* and the *xylS* gene. Deletion of the C-terminal helix-turn-helix motif from NifA protein reportedly did not eliminate NifA-dependent activation of the *nifH* promoter (*23*). Therefore, NifA protein seems to have an ability to directly contact the NtrA-containing RNA polymerase bound at the promoter. The role of the upstream regulatory sequence would be to increase the local concentration of NifA protein near the target promoter. Protein-protein interaction between the activator and NtrA-containing RNA polymerase would be most important in this activation mechanism. The URSs of *OP1* and the *xylS* gene would also facilitate this protein-protein interaction by providing XylR protein to NtrA-containing RNA polymerase on the promoter.

IX. INVOLVEMENT OF IHF IN TRANSCRIPTION REGULATION

In the deletion analysis of the upstream region of *OP1*, we noticed that the 19-bp segment between −43 and −61 could have a role in efficient induction (*13*). The integration host factor (IHF) is involved in NtrA-dependent activation systems (*24*). The 19-bp segment of *OP1* contained the consensus IHF-binding sequence. In an *E. coli* IHF-mutant, which lacks both the α and β subunits of IHF, the activation of *OP1* in the presence of xylene was abolished (*25*). Therefore, IHF was considered to be a factor necessary for the efficient activation of *OP1*. The identification of the IHF-binding site in the upstream region of *OP1* has recently been reported (*15*). It was demonstrated that IHF induces a bend in the upstream region of *nifH* and *glnA* promoters (*24*).

In contrast to the reduced expression of *OP1* in the *E.coli* IHF-mutant, induction of the *xylS* gene expression in this mutant cell was not significantly impaired (*25*). DNase I-footprinting analysis revealed that IHF binds to the sequence which overlaps the *xylS* promoter with an affinity much lower than binds to the upstream region of *OP1*. In *E. coli* cells producing a large amount of IHF from an expression plasmid, the *xylS* expression was greatly reduced, indicating that IHF acts as a repressor rather than an activator in this system. Such a partial repression of the *xylS* gene is consistent with the previous observation that the uninduced levels of XylS protein are extremely low in the cells (*5*). When the *xylS* gene is committed to the active induced transcription, the quaternary complex of URS, XylR, NtrA-containing RNA polymerase and the promoter would be formed after releasing the bound IHF from the promoter region (Fig. 5). Therefore, IHF is not thought to serve as a stabilizer of the DNA-loop in the case of the *xylS* gene. A mechanism other than one using IHF should be necessary for stabilizing the loop structure in the upstream region of the *xylS* gene.

SUMMARY

The TOL plasmid of *P. putida* mt-2 encodes enzymes for catabolism of xylenes and toluene. The genes for the catabolic pathway are organized in two operons, the upper pathway operon and the lower pathway operon, in addition to two regulatory genes, *xylR* and *xylS*.

Both the upper operon and the *xylS* gene require RNA polymerase containing σ^{54} (NtrA, RpoN) for transcription. *Cis*-acting upstream regulatory sequences of 40-bp palindromes were identified *ca* 150 bp upstream from the respective transcription start sites. These sequences serve as the binding sites of the XylR activator. The central region of XylR is homologous to other activator proteins which require σ^{54} and is regarded as the site of interaction with σ^{54}-containing RNA polymerase. Helical periodicity of DNA for maximal activation was observed in the region between the upstream regulatory sequence and the promoter. These results suggest that there is a DNA-loop which tethers the upstream regulatory sequence to the promoter through interaction between XylR protein and σ^{54}-RNA polymerase in the transcriptional activation of the upper operon and the *xylS* gene. Integration host factor is involved in the upper operon activation, but not in the *xylS* gene activation, indicating that the proposed loop structure is stabilized by different mechanisms in these two activation systems.

REFERENCES

1. M.J. Worsey and P.A. Williams: *J. Bacteriol.* **124**, 7-13 (1975).
2. F.C.H. Franklin, P.R. Lehrbach, R. Lurz, B. Rueckert, M. Bagdasarian, and K.N. Timmis: *J. Bacteriol.* **154**, 676-685 (1983).
3. S. Inouye, A. Nakazawa, and T. Nakazawa: *J. Bacteriol.* **155**, 1192-1199 (1983).
4. S. Inouye, A. Nakazawa, and T. Nakazawa: *Proc. Natl. Acad. Sci. U.S.A.* **84**, 5182-5186 (1987).
5. S. Inouye, A. Nakazawa, and T. Nakazawa: *J. Bacteriol.* **169**, 3587-3592 (1987).
6. J-L. Ramos, N. Mermod, and K.N. Timmis: *Mol. Microbiol.* **1**, 293-300 (1987).
7. G.N. Gussin, C.W. Ronson, and F.M. Ausubel: *Annu. Rev. Genet.* **20**, 567-591 (1986).
8. R. Dixon: *Mol. Gen. Genet.* **203**, 129-136 (1986).
9. S. Inouye, A. Nakazawa, and T. Nakazawa: *Gene* **84**, 145-152 (1989).
10. S. Inouye, M. Kimoto, A. Nakazawa, and T. Nakazawa: *Mol. Gen. Genet.* **221**, 295-298 (1990).
11. T. Köhler, S. Harayama, J-L. Ramos, and K.N. Timmis: *J. Bacteriol.* **171**, 4326-4333 (1989).
12. S. Kustu, E. Santero, J. Keener, D. Popham, and D. Weiss:*Microbiol. Rev.* **53**, 367-376 (1989).
13. S. Inouye, M. Gomada, U.M.X. Sangodkar, A. Nakazawa, and T. Nakazawa: *J. Mol. Biol.* **216**, 251-260 (1990).
14. M. Gomada, S. Inouye, H. Imaishi, A. Nakazawa, and T. Nakazawa: *Mol. Gen. Genet.* **233**, 419-426 (1992).
15. V. de Lorenzo, M. Herrero, M. Metzke, and K.N. Timmis: *EMBO J.* **10**, 1159-1167 (1991).
16. S. Inouye, A. Nakazawa, and T. Nakazawa: *Gene* **66**, 301-306 (1988).
17. M. Drummond, P. Whitty, and J. Wootton: *EMBO J.* **5**, 441-447 (1986).
18. S. Inouye, M. Yamada, A. Nakazawa, and T. Nakazawa: *Gene* **85**, 145-152 (1989).
19. M. Merrick, J. Gibbins, and A. Toukdarian: *Mol. Gen. Genet.* **210**, 323-330 (1987).

20. E. Morett and M. Buck: *J. Mol. Biol.* **210**, 65–77 (1989).
21. D.L. Popham, D. Szeto, J. Keener, and S. Kustu: *Science* **243**,629–635 (1989).
22. W. Su, S. Porter, S. Kustu, and H. Echols: *Proc. Natl. Acad. Sci. U.S.A.* **87**, 5504–5508 (1990).
23. E. Huala and F.M. Ausubel: *J. Bacteriol.* **171**, 3354–3365 (1989).
24. T.R. Hoover, E. Santero, S. Porter, and S. Kustu: *Cell* **63**, 11–22 (1990).
25. H. Imaishi, M. Gomada, S. Inouye, A. Nakazawa, and T. Nakazawa: in preparation.

Molecular Structure and Life (Ed. by Y. Kyogoku and Y. Nishimura), pp. 261–272, Japan Sci. Soc. Press, Tokyo/CRC Press, Boca Raton (1992)

DNA Looping and Gene Regulation

Robert SCHLEIF

*Biology Department, Johns Hopkins University**

In general cells are self-contained economies. They contain a collection of factories or machines synthesizing parts that are then used in the construction of these very same machines as well as in synthesizing structural parts of the cells. Many strains of bacteria are totally self-contained, requiring only the import of raw materials and energy. Most higher cells are not totally self-contained, but import a few relatively simple parts which they easily acquire from their environment and whose synthesis would require much enzymatic machinery or use much energy. Such independent biochemical economies are highly complicated, even in bacteria, and therefore a tremendous amount of information is required for their proper functioning. The flow of this information from its storage in the DNA to the parts of the cell that need and use the information must be regulated. This review concerns an unanticipated aspect to the regulation of the flow of this information. Before explaining the biological situation, it is helpful to consider an analogous, but much simpler situation in a computer.

Large amounts of information are stored on magnetic tapes for

* Baltimore, MD 21218, U.S.A.

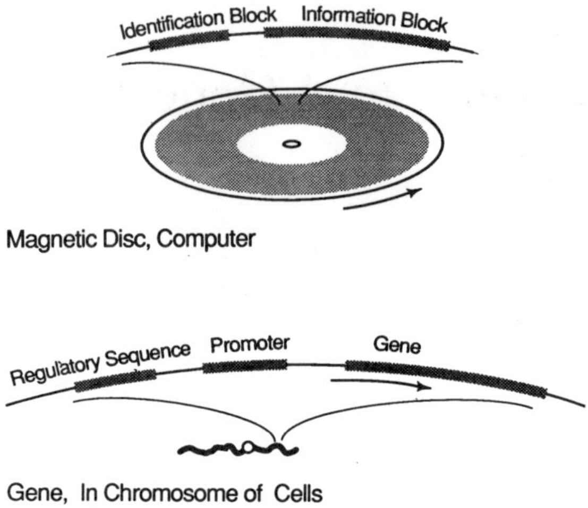

Fig. 1. Structure of stored data on a computer magnetic disc and structure of stored information in the chromosomes of cells. In both, identification blocks precede the information blocks.

computers, on floppy discs, and on hard discs. As in the cell, not all this information can be read out at once, as there are insufficient reading devices. Furthermore, it is of no value to the computer programs to be flooded with irrelevant information. In a computer, as in a cell, the flow of information from a storage location to its site of need, must be precisely regulated. When a block of information stored on magnetic medium is needed, a reading device looks for the information. How is the required information identified? We cannot merely look for it by seeking the information itself, for this would require knowing beforehand the information itself. Instead, the information is stored in blocks, and each block is preceded by an identification area (Fig. 1). The computer program asks for the information from block x, and the readout mechanisms then seek and ultimately read out the information stored adjacent to the identifier x.

Information storage and retrieval in cells works the same way as in computers. Earlier work in molecular genetics has shown that cellular information is stored in the DNA in blocks called genes. The identification blocks that are used by the readout machinery are promoters and the binding sites of regulatory proteins. Early experiments in genetics and biochemistry found that these identification and readout-assisting blocks or sequences were immediately adjacent to the information blocks (Fig. 1). We can call such an arrangement a local information identification

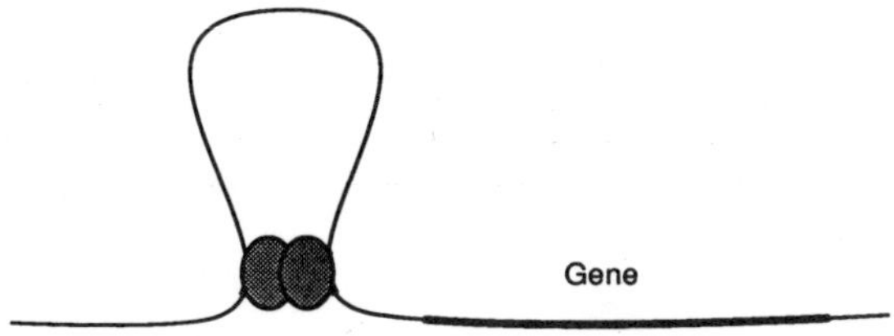

Fig. 2. DNA looping in which both a sequence block adjacent to a gene and a sequence block at some distance from a gene are linked by a bridging protein. By this means more distal DNA sequences may also participate in the regulation of genes.

scheme. That is, the true stored information is immediately beside the identifiers that are used in the identification and readout of the information. The remainder of this review concerns the unexpected discovery that cells use both a local identification scheme, and also a nonlocal identification scheme. More precisely, the nonlocal modification should be called a semi-local identification scheme because it does identify information in the vicinity, but that information may not be immediately adjacent. In order to utilize the local and semi-local information, a readout or identification complex must simultaneously read both sequences at once. Thus, the nonadjacent sites must be part of a complex, and hence, the DNA between the sites is looped by the proteins binding to the two sequences. As mentioned above, the existence and utilization of such loops was unexpected. They were first found in the arabinose system in *Escherichia coli* where they have been extensively characterized (*1–7*). With the techniques developed for study of the arabinose system and the clear biochemical demonstration of such loops, they subsequently have been found to occur in most regulated gene systems in all types of organisms.

I. THE ARABINOSE GENE SYSTEM AND DNA LOOPING

The first gene systems carefully studied were the lactose operon of *E. coli* and phage lambda. In both of these a repressor protein was found to bind close to the regulated genes and to prevent transcriptional readout of the adjacent gene. Early experiments by Ellis Englesberg indicated that the L-arabinose utilization system in *E. coli* was different (*8*, *9*). Here, the regulatory protein that senses the presence of arabinose acts positively to turn on the synthesis of the proteins necessary for the utilization of arabinose instead of negatively as in the *lac* and lambda systems. This

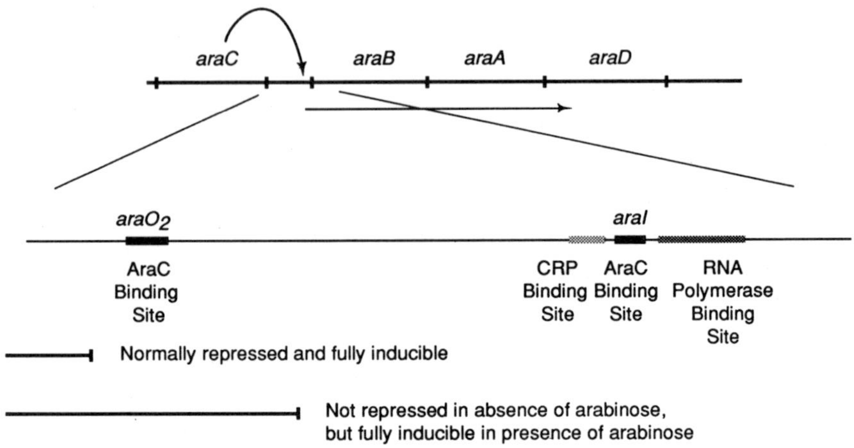

Fig. 3. Top, structure of the arabinose operon. The AraC gene protein acts in the presence of arabinose to induce transcription of the *araB*, *araA*, and *araD* genes. The regulatory region of the genes lies between *araC* and *araB*. Middle, expanded view of the 300 base pair regulatory region showing the binding sites for AraC protein, CRP protein, and RNA polymerase. Bottom, deletions entering the arabinose operon from the left and ending before $araO_2$ are normally repressed in the absence of arabinose and fully inducible in the presence of arabinose. Deletions that extend beyond $araO_2$ are not repressed in the absence of arabinose and have elevated basal levels.

fact stimulated interest in the system and indicated to biochemists that nature might not be so simple as to use one basic mechanism for the regulation of all genes.

Further study of the arabinose system indicated that in addition to its positive component in gene regulation, the regulator AraC protein also acted negatively to repress (*10*). This negative aspect to the regulation was revealed by a deletion of material from upstream of the *araBAD* genes. The *araBAD* genes remained fully inducible, indicating that the identifying sequences and readout sequences were intact. Nonetheless, regulation of the *araBAD* genes was altered. Their uninduced or basal level was raised. A variety of relatively trivial explanations for this elevation were eventually eliminated, and the effect was shown to be mediated by AraC protein itself. As the deletion entered the arabinose operon from upstream of the *araBAD* genes, it showed that some sequence or site that was not essential for induction, but which was necessary for negative regulation by AraC protein was located upstream from all the sites necessary for induction. Of course, upstream is not a quantitative term. The genetics experiments could not locate this upstream site.

Over the years our laboratory has directed much of its work toward understanding this repression from upstream phenomenon. One of the first steps in the study was to repeat the work of Englesberg. In a careful and extensive deletion analysis of the system, we found a sizeable number of deletions that had the same propery as the two originally isolated by Englesberg (*11–13*). Within other parts of the arabinose operon the fraction of the total deletions that ended in an interval was found to be roughly proportional to the size of the interval. This indicated that an approximate measure of the distance to the upstream site required for repression could be obtained by consideration of the fraction of the deletions ending in the arabinose operon that had lost the repression abilities but still retained normal induction properties. This fraction indicated that the site required for repression was located at least several hundred base pairs upstream from all the sites required for induction. Such a separation between a regulatory site and the regulated gene was unexpected, and at that time not even believed.

The developments of genetic engineering were required to locate more precisely the upstream site involved in repression of the arabinose operon. New deletions lacking the repression site were then isolated and their deletion endpoints were determined by DNA sequencing (*3*). These located the site required for repression at about 200 base pairs upstream from all the sequences required for induction (*1*). Biochemical footprinting *in vitro* indeed located a previously unnoticed binding site for AraC protein in about the expected position 200 base pairs upstream from the sites required for induction. Thus it appeared that AraC protein somehow could act from more than 200 base pairs upstream from the induction site to effect repression on the *araBAD* genes. At the time such an action at a distance phenomenon was not yet known in other systems.

DNA looping was the most plausible, to me, of the several basic mechanisms that suggested themselves for biochemically plausible schemes by which a protein bound to an upstream site could influence proteins bound to DNA several hundred base pairs away. One virtue of this mechanism was that it was experimentally testable. In DNA looping two proteins must bind to each of the two sites involved and bind to each other. The actual pathway for loop formation can be varied. For example, one subunit of a dimeric protein could bind to one site followed by the other subunit binding to the second site only after Brownian motion moved the second site into the vicinity of the second subunit.

Looping could be facilitated or hindered by the orientation of the sites involved. That is, looping would be hindered if the DNA not only had to bend, but if the DNA had to twist so as to place both proteins on the proper side of it. Torsionally twisting DNA is energetically expensive for short stretches of DNA. It costs about 2 kcal/mole to introduce a half a turn of twist into a stretch of DNA 200 base pairs long (*14*). This is a substantial amount of energy, and therefore looping might be prevented merely by requiring such twisting to occur. That is, the introduction of five base pairs of DNA, half a helical twist, at an irrelevant location between the two AraC protein binding sites might block looping and therefore interfere with repression in the system. This is precisely what was found! Furthermore, the additional insertion of five more base pairs, for a total of ten base pairs or nearly one full turn, restored repression. A series of these spacing mutations which showed oscillating repression and lack of repression indicated that, indeed, the mechanism of repression from the upstream site was via DNA looping (*1*). A number of other experiments have since confirmed these initial experiments.

One amusing sidelight to the study of DNA looping was that the helical twist experiments permitted a measurement of the helical repeat of DNA *in vivo*. An extensive set of spacing mutations was measured and the helical repeat was found to be a little over 11 base pairs per turn (*6*). This is different from the values measured *in vitro* on linear DNA of about 10. 5 base pairs per turn. The difference results from the fact that *in vivo* the DNA possesses a linking number deficit that tends to unwind the DNA and therefore increase the number of base pairs per turn. The best estimates of the amount of untwisting suggest that 11 base pairs per turn may be a bit larger than expected, so that factors specific to the arabinose operon or the cells used may also have a small influence on the values found.

The looping and helical twist experiments permitted investigation of the upper and lower limits to loop sizes. As the size of a loop is increased,

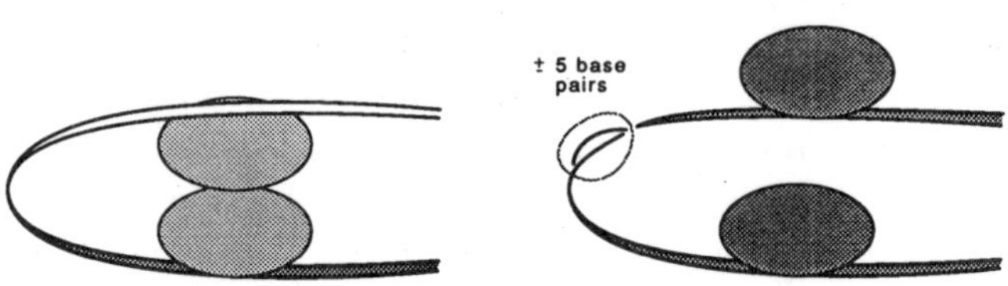

Fig. 4. Introduction of half a helical turn between two binding sites can interfere with DNA looping.

we might expect the probability of looping to fall. This would result from the decreased concentration of one of the sites involved in the looping in the vicinity of the other site. If the distance between the sites went to infinity, there would be no connection between the sites and therefore no increase in the concentration of one site near the other due to the tethering provided by the DNA. In this case the protein would occupy the sites independently of one another, whereas when the sites are separated by a few hundred base pairs of the DNA, the binding of a protein to one site places it near the other site and therefore increases the chances that a second DNA binding site on the protein can bind to the second DNA site. As the spacing between *ara*O_2 and *araI* is increased from the natural size of 211 base pairs, repression of the *araBAD* genes gradually falls, and is undetectable with a spacing of about 450 base pairs (*6*). Such a rapid fall-off in the repression was unexpected, and its origin is not yet understood.

The loops could also be made smaller than the wild type 211 base pairs. If AraC protein were inflexible, we might expect that the formation of small circles would become energetically too expensive and the formation of small circles would be impossible. Curiously, this was not found. Looping was observed, still with a periodicity of about 11 base pairs per turn right down to 30 base pairs separating *ara*O_2 and *araI*. Such a finding indicates that AraC protein is highly flexible (*6*).

In vivo footprinting to examine the occupancy of the *ara*O_2 and *araI* sites provided additional information about looping in the arabinose system (*4*). The first point concerns the *araI* site. Until the discovery of looping in the arabinose system, the general picture of gene regulation was that the expression of genes was turned on or turned off by the binding or dissociation of proteins from DNA sites. This was the expectation for the arabinose system as well. From *in vitro* studies it was known that arabinose increased the affinity of AraC protein for the *araI* site by about 30 fold. Thus, it was thought that upon the addition of arabinose, AraC protein bound to the *araI* site, and from there stimulated transcription of *araBAD*. On the contrary, the data indicating looping in the arabinose system showed that deletion of the *ara*O_2 site resulted in the increase in basal level expression or the loss of repression. As this occurred in the absence of arabinose, it indicated that looping normally existed and involved the *ara*O_2 site and something else in the regulatory region. The most likely second site to be involved in looping was AraC protein bound at the *araI* site. That is, if looping were true, AraC protein should occupy the *araI* site even in the absence of arabinose. A method for *in vivo* footprinting was therefore developed and applied to the question of whether the *araI* site was indeed occupied in the absence of arabinose (*4*). It was. Not only did this support DNA looping, but it also meant that genes need not be regulated exclusively by the binding or release of proteins from DNA. Instead, genes can also be regulated by changes in conformation of proteins bound to DNA.

Application of *in vivo* footprinting to the *ara*O_2 site provided a very nice confirmation of DNA looping. As mentioned above, the ability to loop between two sites generates cooperativity in the binding of a protein to the two sites. This cooperativity is pronounced in the *ara* system. The affinity of AraC protein for the *ara*O_2 site is 30 fold lower than its affinity for the *araI* site. At the concentrations of AraC protein found *in vivo*, the *araI* site is normally occupied, and in wild type cells the *ara*O_2 site is also occupied by AraC protein. When, however, the *araI* site region is deleted, the remaining *ara*O_2 site is no longer occupied by AraC protein (*4*). This demonstrates the cooperativity in the looping between *araI* and *ara*O_2.

Recent looping experiments on the arabinose system have provided a possible mechanism for the regulation of loop formation and dissociation (*7*). While looping appears to be necessary for repression in the arabinose system, the available evidence indicates that looping is not necessary for induction in this system. Two clues hint at this. First, normal induction is possible in strains deleted of the *ara*O_2 site. Thus, with only the *araI* site present, induction is still possible. Second, the *in vivo* footprinting experiments show that immediately after the addition of arabinose the occupancy of the *ara*O_2 site falls. These clues indicate that upon the addition of arabinose the loop opens and induction follows. How might the loop be opened, and how could we show this?

The entropic and energetic costs of forming DNA loops can be large. Since the strength of the looping interactions in the arabinose system is not high, study of looping *in vitro* was hampered until a nice system for looping was developed in the *lac* system (*15*). When the sites involved in looping are placed in a small supercoiled DNA circle, the structure of the supercoiled DNA has already brought the sites into close proximity so that little additional energy is required to bridge the two sites and form a loop. In the *lac* and in the *ara* systems such looped small circles migrate at a faster rate than the free DNA or DNA on which the protein is bound at only one site (*15*). Presumably this faster migration rate results from the protein holding the structure in a highly compact form. The faster migrating species was formed in the arabinose system only when both the *araI* and the *ara*O_2 sites were present. It was absent when one of the sites was rotated to the back face of the DNA by the introduction of five base pairs on either site of the site. Further, the stability of the complex was greater than the stability of the species formed by binding to either the *araI* site alone or the *ara*O_2 site alone. Thus, this system forms a convenient assay for looping.

As expected from the comments above, when arabinose was added to a looped complex, the loops were broken. This opening of the loops is an intrinsic property of the protein. It does not depend upon the identity of the two sites *ara*O_2 and *araI*. These sites, while similar, are not identical, and therefore we might have imagined that the protein could tell the two apart and in the presence of arabinose might be built so as to let go of the *ara*O_2 site. This is not the way the system works, however, for a loop formed between two *araI* sites opens when arabinose is added.

AraC protein is a dimer in solution and it binds to the *araI* site on linear DNA as a dimer (*16*). Therefore, it was expected that the loop would be formed from the coming together of a dimer of the protein bound at *araI* and a dimer bound at *ara*O_2 to form a tetramer. This is not what is found. The first clue that the looped species might be something different was the finding that a dimer of AraC could be bound at the *araI* site in the presence of arabinose. Then, free protein could be removed from the solution, and then arabinose removed. The previously unlooped protein-DNA complexes formed loops. That is, the dimers bound at the *araI* site changed their state and then bound to both *araI* and *ara*O_2. The most plausible way this could occur is that the looped species is formed from a dimer, with one monomer binding to half the *araI* site and the second monomer contacting the *ara*O_2 site (*7*).

If the looped species is indeed formed from a dimer and only half the *araI* site is contacted by AraC protein, several results should follow. First, premodification experiments should reveal which part of the *araI* site is not involved with looping, and second, if we change the sequence of the uncontacted part of the *araI* site, then looping should not be affected, but

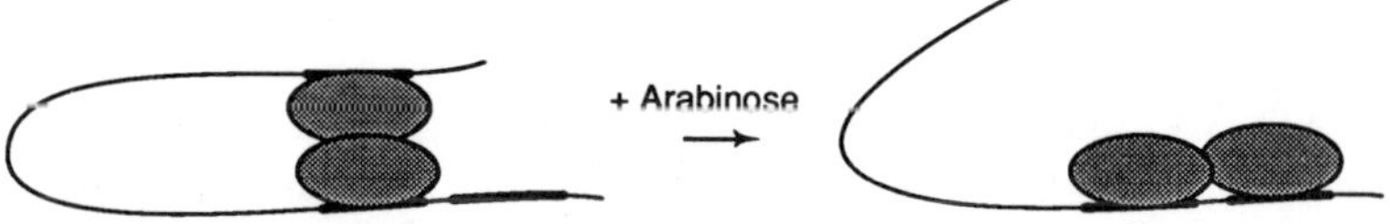

Fig. 5. Upon the addition of arabinose, the two subunits of AraC protein prefer to bind to two adjacent sites rather than two nonadjacent sites.

binding of AraC in the unlooped state should be greatly weakened. Both of these were observed (7). In premodification experiments the DNA was methylated on guanine residues at random and then those molecules still capable of forming loops and those molecules incapable of forming loops were separated by electrophoresis. Those guanines whose methylation interfered with looping were identified as lying in the right half of the *araI* site, and those essential for looping were in the left half of the *araI* site. Accordingly, a modified *araI* site was synthesized retaining the normal sequence in the left half and severely modified in the right half. This DNA was fully capable of forming DNA loops. Furthermore, their stability as assayed by their dissociation time was the same as loops formed from the fully wild type *araI* site. Upon arabinose addition, the loops on the modified DNA disappeared, and AraC was incapable of binding to the DNA altogether. These experiments showed that it is a dimer of the protein which forms the loop, and that upon the addition of arabinose the dimeric protein shifts its contacts from the $araO_2$ site to the *araI* site. Direct stoichiometric experiments confirmed that it is a dimer of the protein that forms the loops (7).

By what mechanism could the protein prefer to form loops in the absence of arabinose and prefer to bind to two adjacent half sites in the presence of arabinose. One of the simplest ideas is that the subunits are oriented in the absence of arabinose such that they would have to be significantly distorted to contact the two adjacent half sites on DNA. Thus, it would be energetically more favorable for the protein to form a loop in the DNA with one subunit contacting half of the *araI* site and the other contacting the $araO_2$ site. In the presence of arabinose the subunits could reorient such that they were properly positioned for binding to two adjacent halves of the *araI* site. Therefore, the addition of arabinose would open the looped form. Furthermore, on linear DNA the presence of arabinose would increase the binding affinity because most of the binding energy could go into holding the protein on the DNA instead of using part of the binding energy to force the subunits into the correct orientation.

SUMMARY

Why is DNA looping used in biological systems? Two answers appear to exist to this question. First, DNA looping permits multiple proteins to interact with an active complex on the DNA, for example, an RNA polymerase molecule bound at the promoter of a gene. By virtue of looping not only can regulatory proteins bind alongside the RNA polymerase to affect its activity, but they can also bind at sites several hundred or several thousand base pairs away and interact with the complex through DNA looping. Looping permits more complicated systems to operate. By analogy to the computer systems, they enlarge the identification regions for blocks of information.

The second reason for DNA looping is the cooperativity of the protein binding. In the nucleus of the cell many thousands of genes must be regulated. Therefore, many thousands of regulatory proteins must be present. As a result of this large number, no protein can be present in very high concentration. If the protein is to bind to regulatory sites, its affinity for the site must be high. Attaching a protein to a site with such high affinity could generate problems for the cell. Replication and repair processes could be impeded by the presence of the bound protein. A simple way to solve this problem is to utilize the cooperativity inherent in DNA looping. When looping is possible the actual affinity of a protein for a binding site can be significantly lower, and through the cooperativity, or equivalently, the increase in the concentration of the protein at one site due to its binding to the other site, the site can still remain occupied most of the time.

In summary, DNA looping was an unexpected mechanism to be involved in regulating transactions among DNA binding protein complexes, and yet on further analysis, its use in nature is understandable.

Acknowledgments
The work from my laboratory has been supported on NIH grant GM18277. I thank the many people in my laboratory who, over the years, have worked with me in studying gene regulation and the arabinose operon.

REFERENCES

1. T. Dunn, S. Hahn, S. Ogden, and R. Schleif: *Proc. Natl. Acad. Sci. U.S.A.* **81**, 5017–5020 (1984).
2. S. Hahn, T. Dunn, and R. Schleif: *J. Mol. Biol.* **180**, 61–72 (1984).
3. T. Dunn and R. Schleif: *J. Mol. Biol.* **180**, 201–204 (1984).
4. K. Martin, L. Huo, and R. Schleif: *Proc. Natl. Acad. Sci. U.S.A.* **83**, 3654–3658 (1986).
5. L. Huo, K. Martin, and R. Schleif: *Proc. Natl. Acad. Sci. U.S.A.* **85**, 5444–5448 (1988).
6. D. Lee and R. Schleif: *Proc. Natl. Acad. Sci. U.S.A.* **86**, 476–480 (1989).
7. R. Lobell and R. Schleif: *Science* **250**, 528–532 (1990).
8. E. Englesberg, J. Irr, J. Power, and N. Lee: *J. Bacteriol.* **90**, 946–957 (1965).

9. D. Sheppard and E. Englesberg: *J. Mol. Biol.* **25**, 443–454 (1967).
10. E. Englesberg, C. Squires, and F. Meronk: *Proc. Natl. Acad. Sci. U.S.A.* **62**, 1100–1107 (1969).
11. J. Lis and R. Schleif: *J. Mol. Biol.* **95**, 395–407 (1975).
12. R. Schleif and J. Lis: *J. Mol. Biol.* **95**, 417–431 (1975).
13. R. Schleif and J. Lis: *Proc. Natl. Acad. Sci. U.S.A.* **69**, 3479–3484 (1972).
14. D. Shore and R. Baldwin: *J. Mol. Biol.* **170**, 957–981 (1983).
15. H. Krämer, M. Amouyal, A. Nordheim, and B. Müller-Hill: *EMBO J.* **7**, 547–556 (1988).
16. W. Hendrickson and R. Schleif: *Proc. Natl. Acad. Sci. U.S.A.* **82**, 3129–3133 (1985).

Subject Index